100 DESIGNS/100 YEAR

100 DESIGNS / 100 YEARS

INNOVATIVE DESIGNS OF THE 20TH CENTURY
MEL BYARS WITH ARLETTE BARRÉ-DESPOND

RotoVision

Published by RotoVision SA
Rue du Bugnon, 7
CH-1299 Crans-Près-Céligny
Switzerland

RotoVision SA
Sales & Editorial Office
Sheridan House
112/116A Western Road
Hove BN3 IDD, England
Tel: +44–1273–7272–68
Fax: +44–1273–7272–69
E-mail: sales@RotoVision.com

ISBN 2-88046-442-0

10 9 8 7 6 5 4 3 2 1

This book was written, designed, and produced
by Mel Byars.

Printed in Singapore
Production and separation by
ProVision Pte Ltd, Singapore
Tel: +65–334–7720
Fax: +65–334–7721
E-mail: prov@pacific.net.sg

Contents

continued >

The Fifties

1950 Armchair by Ernest Race
1951 Side chair by Gio Ponti
1952 Lounge chair by Harry Bertoia
1953 Floor lamp by Serge Mouille
1954 Ballpoint pen by Don Doman and others
1955 Decanters by Timo Sarpaneva
1956 Phonograph-radio by Hans Gugelot and Dieter Rams
1957 Sewing machine by Marcello Nizzoli
1958 Club chair by Arne Jacobsen
1959 Dinnerware by Hans Roericht

The Sixties

1960 Stacking chair by Verner Panton
1961 Electric typewriter by Eliot Noyes
1962 Electric razor by Gugelot, Rams, and Müller
1963 Scissors by Olaf Bäckström
1964 Radio by Marco Zanuso and Richard Sapper
1965 Stacking chair by Cesare "Joe" Colombo
1966 Clock by Gino Valle
1967 Armchair by De Pas, D'Urbino, Lomazzi, and Scolari
1968 Chaise longue by Olivier Mourgue
1969 Seat by Gatti, Paolini, and Teodoro

The Seventies

1970 Chest by Shiro Kuramata
1971 Table lamp by Richard Sapper
1972 Camera by Henry Dreyfuss and Associates
1973 Food processor by Carl Sontheimer
1974 Stereo tape deck by Mario and Dario Bellini

1975 Chair by Gaetano Pesce
1976 Bicycle lock by Michael S. Zane
1977 Picnic set by Jean-Pierre Vitrac
1978 Flatware by Donald Wallance
1979 Portable radio-cassette unit by the Sony Design Team

The Eighties

1980 Adhesive note paper by Arthur Fry and Spence Silver
1981 Bookcase by Ettore Sottsass
1982 Portable computer by I.D. Two
1983 Wristwatch
1984 Computer mouse by Hartmut Esslinger and frogdesign
1985 Kettle by Michael Graves
1986 Table by Foster and Partners
1987 Computer by Hartmut Esslinger and frogdesign
1988 Armchair by Jorge Pensí
1989 Lamp by Alberto Meda and Paolo Rizzatto

The Nineties

1990 Lemon press by Philippe Starck
1991 Chest by Tejo Remy
1992 Armchair by William Stumpf and Donald Chadwick
1993 Razor by Ross Lovegrove
1994 Chest by Antonio Citterio with G. Oliver Löw
1995 Chair by Marcel Wanders
1996 Lamp/table/stool by Tom Dixon
1997 Mixing bowl by Hansjerg Maier-Aichen
1998 Computer by Jonathan Ive and others
1999 Vacuum cleaner by Ljunggren, Ljunggren, and
     Haegermark

Queen for a Day

By Arlette Barré-Despond

When Mel Byars and I thought of doing a book that would present a view of the past century, we began with the idea that each of us would separately select our "objects of the century." Then we would compare our choices and arrive at a common approach to what the design of the century is. And with Mr. Byars in New York and me in Paris, a "transatlantic vision" would hopefully emerge. This has happened. But the truth is that Mel Byars selected the entries, and, in most cases, I approved of them.

To open the book with a picture of the "Brownie" camera by George Eastman was his idea. When we measure today the importance and status of manmade images—from the tangible to the virtual—this particular choice has also become mine. However, I would probably not have spontaneously chosen it myself. Of course, any one of many other objects could have occupied first place. And one or another of them would have been more revealing of the state-of-thought at the dawn of the century—objects steeped in historicism against which the first designers would eventually lead an unmerciful battle.

Wherever you live, in the United States or Europe, it is very easy to establish that this century found its inspiration in historic styles. Only a few visionaries attempted to conceive a new art—an "Art Nouveau" as it came to be known—in which historians like Nikolaus Pevsner saw the outset of modernity. The Art Nouveau style reached its acme in 1900 when it became discredited due to a great number of stylistic extravagances. These excesses lead other innovators and visionaries, like Le Corbusier in particular, to specifically pose: What will be the style of the century to come?

The past has been distinctly marked by the powerful will of monarchs who imposed their own style. This can be seen, for example, in the differences between Louis XIII and Louis XVI. But the 20th century took its own time in revealing its own characteristics. And modernity could not be separated from the idea of the emancipation of people. ("Modernity" should not be confused with the stylistic, commercial caricature of "Modernism.")

Only after World War I, when the question of ornamentation had been momentarily closed, was there the advent of the concretization of the long-contained aspirations for more justice and a world where production would serve the needs of the underprivileged. So what was the mission of the creators, architects, and designers? Most of them, who are well represented here, answered the calling to serve daily life, domestic life. And daily objects, domestic objects, such as chairs, kettles, lighting fixtures, cameras, and wristwatches, which had been in the past reserved for a certain élite were becoming democratized, and sometimes even disposable.

100 Designs/100 Years shows one possible history of design. Anyone can write his own. The chosen images, the objects, and the accompanying text provide a subjective reading from two design historians in love with the work they do.

At the end of a century rich in spectacular overhangings and significant withdrawals, it is now possible to make a rough sketch of its many stages. Aside from the opening image of Eastman's camera, which places the century under the sign of modernity and the entries at the end which have appropriated élitist practices, the history of 20th-century design can now be told.

By the mid-1920s, a modernity at odds with historical models had evolved. There were the avant-garde voices and manifestoes of the Vienna Secession and Josef Olbrich, Czech Cubism and Josef Chochol, Russian Constructivism and Kazimir Malevich, A.E.G. and Peter Behrens, and the seminal style of Frank Lloyd Wright, as well as icons like the Coca-Cola bottle and Gerrit Rietveld's armchair, a child of De Stijl.

During this time, Europe was the model observed and copied. In spite of the 1907 immigration laws, many European creators such as Paul Frankl, John Vassos, Eliel Saarinen, and Kem Weber settled in America. From the time of the 1925 and 1937 Paris expositions to World War II, modernity flaunted itself. Forms were refined, and endless reproductions were possible. There were Le Corbusier, the Bauhaus, Ludwig Mies van der Rohe, Raymond Loewy, Alvar Aalto, and daily life was being revisited with the "Zippo," the "Moka Express," the first Rayban eyeglasses, and the people's radio, symbolic of Nazi production. All these people and creations were exciting, but the times were not. The rise of European fascism, the Wall Street crash in the U.S., and other disturbances were emerging. The "Barcelona" chair by Mies, the pencil sharpener by Loewy, the "Hermes-Baby" typewriter by Giuseppe Prezioso, the "Savoy" vase by Aalto, and other icons were checking in. American capitalism was becoming efficient, and its firms were calling on professional designers in droves. Yet Europeans did not likewise respond. "Tupperware" by Earl Tupper was spreading over the entire world as was "Scotch" tape,

Wurlitzer jukeboxes, and Eames chairs.

In the 1950s and '60s, new materials mainly derived from petroleum encouraged the emergence of new forms. Military industries turned to the service of peacetime production. More a secret up to this moment, Scandinavia began exporting its know-how that amalgamated industry and craft. Italy, Japan, and Germany, the defeated countries of the war, were on the way to becoming the most representative design communities in the world. Their industries had recovered, and talented young and not-so-young designers were at work. Italy and Germany are well represented on these pages for different reasons—Italy for its incredible and still-ongoing creativity and Germany for its perfect attachment to functionalism.

The 1970s, '80s, and '90s saw the gasoline crisis, the dilemma of functionalism, the rise of Postmodernism, the return of ornamentation and the baroque, the introduction of new technologies and composite materials, the question of morality and ethics in design, and numerous other issues. Everything cannot be said on these pages.

While, of course, other books exist on design history, the outline in this one has been clearly unraveled, written, and shown. One fact is sure: objects are the testimony of who we are, speak of our quest for identity, and reveal how we live.

One Year, One Design

By Mel Byars

An honest history of 20th-century design must necessarily be told not only through the lives of objects but also those of their creators. Yet, history aside, objects provoke immediate emotions all by themselves. They can be bigger than life with a recognition beyond that of their creators and makers. But history, as any school child can confirm, is often dry. An image is immediate and powerful, while books and video programs about history demand time and patience, unless you just look at the pretty pictures. The feel and look of objects are far more stimulating and intriguing than dry technical and intellectual concerns and historical information. Nevertheless, there is a great deal of history in this volume.

The objects in the survey range from the anonymous to the élite. The anonymous ones tend to be the ordinary tools of life, ones we cannot do without or even want to. Most of us never know or care who designed them. We might not ever consider the ardor and intelligence behind their creation. Even so, with some patience, their histories might be highly entertaining.

The élitist ones may not be élitist at all. Unfortunately, their placement in the pantheon of design-museum collections and the almost endless effusing about them by historians and journalists may have ironically given them a bad reputation. The work of trained, experienced designers who have a knowledge of what has passed and an insight into what might come may be the most qualified to give us what we want, even though we may not know what we want. Such is the nature of the capitalist marketplace.

The assignment of specific dates to the objects here has been made possible in part by the advent of the fairly new discipline of design history, from the records of some designers, and with the cooperation of some manufacturers. Unfortunately, the dates of design or first manufacture published in journals, magazines, and books do not always agree. An assigned design date may be the year when the design process began or when it ended, or an accredited date may simply be incorrect. Yet, great effort was made here to provide correct dates which, even so, should not be accepted as gospel.

Most design specialists only occasionally consider an exact date to be absolutely important. Normally they think of an object as being of a decade or so or within the range of time of a particular movement or style. They can see or find out from colleague-specialists that an object predicted a style or appeared at the threshold of a style's birth, during its effloresce, or in its decline. Otherwise, a precise year may not matter unless, of course, it marks a specific event, such as Sottsass's bookcase in the first collection of Memphis (see p. 174) or Olivetti's typewriter that was first shown at an important exposition (see. p. 32).

A group of particularly distinctive objects that appeared within the same year created a dilemma. Which to choose? Like films vying for the Palm d'or at the festival in Cannes or contenders for a book award, one object was as good, based on the arbitrariness of "good," as the other. I asked the opinions of a number of design experts, historians, and journalists. In some few cases when the pickings within a single year were slim, one certain object was, of course, the unanimous choice. More often than not, their preferences differed from one another. I seriously considered their rationale and embraced their inestimably valuable advice and chose the final entrants myself.

The idea for the book—one object for each year of the 20th century—was design historian Arlette Barré-Despond's. In the midst of preparing the book, I thought the premise was absurd. How can a single object represent an entire year? But, as with all contests, a winner does not represent all others of the same year; it ultimately represents itself. Looking at the final form, my guess is that the idea was probably a good one.

100 DESIGNS / 100 YEARS

"Brownie" camera by George Eastman

Inventor George Eastman (1854–1932) had a virtue in common with many other American inventors: a fertile imagination married to a lively entrepreneurial spirit. His inventions included a process for making photographic dry plates (1880), flexible film (patented in 1884), and the "Kodak" camera (1888). One of the most distinctive in his line of image-making machines was the "Brownie." The device, in the form of an uncompromisingly simple box, could be operated by a child. In fact, Eastman aimed at making hundreds of thousands of boys and girls into amateur photographers, and he may have succeeded. The name "Brownie" was derived from the series of books by Palmer Cox (1840–1924) about fictional little people called Brownies who became as familiar to the children of 1900 as Mickey Mouse is to children today. The camera was composed of a box configuration which had been established by the 1897 Zar and 1899 Yale cameras, the film-roll holder found in the "Kodak" and "Bulls-Eye" cameras, and a shutter invented by Frederick H. Kelly, formerly of the Blair Camera Co. Even though the very first model was poorly constructed, the camera was a resounding success. Over 50,000,000 were sold for $1.00, with 10¢ extra for paper film or 15¢ extra for transparent film.

Date: 1900. Materials: Cardboard body covered with "leatherette" paper, wood internal reinforcements, metal shutter and winder. Manufacturer: Frank A. Brownell for Eastman Kodak Co., both Rochester, NY, U.S. Photograph: Courtesy George Eastman House, Rochester, NY, U.S.

Events of 1900:

Exposition universelle, which featured the first escalator, opened, Paris (50,000,000 visitors).

Premiers: The auto travel manual ("Guide Rouge Michelin"), France; Hector Guimard's Art Nouveau-style Métro (subway) station entrances, Paris; international lawn-tennis award (the Davis Cup), by D.F. Davis, England; professional wrestling champion (William Muldoon); *Daily Express* newspaper, London; New York auto show; revolvers by Browning; offshore oil wells; rigid airship (dirigible), built and flown by Count von Zeppelin, Germany.

The Cake Walk became the most fashionable dance step.

Giacomo Puccini composed the opera "Tosca," Rome.

Literature: Baum's *The Wonderful Wizard of Oz*, Colette's *Claudine à l'école* (first of the "Claudine" novels), Conrad's *Lord Jim*, Chekhov's play "Uncle Vanya," Heer's *König der Bernina*, Freud's *The Interpretation of Dreams*.

French Impressionist artist Claude Monet first showed *Les nymphéas* (painted 1883), at the Durand Ruel, Paris.

French painter Paul Gauguin's *Noa Noa* reported on his travels in Tahiti.

Société des Artistes Décorateurs was founded by René Guillère and others, Paris. (First Salon exhibition was held at the Petit-Palais in 1904.)

Arthur Evans discovered the Minoan culture, Crete.

Boxers rose up against the Europeans, ending in 1901 with the Peace of Peking, China.

Shintoism was reinstated against the Buddhist influence, Japan.

In about 1880, architecture and design students began rebelling against the prevailing Beaux-Arts style. Claiming to be against its interpretation of historical forms, the style the young architects and designers developed also had earlier roots. They extolled ornamental curves and lines (possibly from Scotland or Austria) and floral forms (possibly from France or Belgium) and combined them with vestiges of the Beaux-Arts. The new style became known as Art Nouveau in France, Jugendstil in Germany, style des Vingt in Belgium, stile liberty in Italy, and by a number other names such as modernismo in Spain and Style 1900. "Art Nouveau" was derived from Siegfried Bing's L'Art Nouveau, a shop in Paris. One of the style's distinguished Austrian exponents and a former student of Wagner and Hoffmann (see pp. 24, 28), Josef Maria Olbrich (1867–1908), was active in both architecture and the decorative arts, including metal. His work was clean and functional compared to much of the florid, undulating expressions of others of the period. Even though he worked in silver, his most popular objects, like the candelabra here in pewter for a smithy near Düsseldorf, were produced in less expensive materials. But all Olbrich objects are scarce today. Unfortunately, he died of leukemia at an early age, only 10 years after his most important commission, the Secession exhibition building in Vienna.

Date: *c.* 1901. Material: Polished pewter. Manufacturer: Eduard Hueck KG, Lüdenscheid, Germany. Photograph: Courtesy Die Neue Sammlung, Munich, Germany, inv. nos. 15/96-a,b. Photo: Angela Bröhan, Munich.

Events of 1901:

Following the so-called century of steam, the century of electricity began.

Wright brothers flew their first glider, U.S.; a motor-driven airplane was first flown, by G. Whitehead.

Marconi received the first transatlantic telegraphic radio transmission (the letter "S"), sent from Cornwall, England, to St. Johns, Newfoundland, Canada.

Premiers: Instant coffee; the Nobel Peace Prize, to Henri Dunant, the Swiss founder of the Red Cross; legal boxing, England; Mercedes auto, by Wilhelm Maybach at Daimler, Germany; suction vacuum cleaner (refrigerator size), by British inventor Booth; "safety" razor with removable blade, patented by Gillette and Nickerson, U.S.; motor-driven bicycles; electric typewriter (Blickensderfer Electric), Germany; oil drilling in Persia (today known as Iran).

The Little Doctor was filmed, England.

Art: Munch's *Girls on the Bridge* and Feradin and Holder's *Spring.* Picasso began his Blue Period, Paris.

Literature: Kipling's *Kim,* Strindberg's "Dance of Death," Couperus's *Babel,* Philippe's *Bubu de Montparnasse,* Norris's *The Octopus,* Planck's *Laws of Radiation.*

Music: Ravel's "Jeux d'eau," Rachmaninov's "Piano Concerto No. 2," and Bruckner's "Symphony No. 6," performed posthumously. Giuseppe Verdi died, Italy.

Adrenaline was isolated and synthesized by Thomas Bell and Japanese biologist Jokichi Takamine, independently.

Havelock Ellis began his *Studies in the Psychology of Sex* (sixth and last volume in 1910), U.S.

Trans-Siberian Railroad reached Port Arthur, China.

First British submarine was launched.

Archibald Knox (1864–1933), after studying art from 1878–84 on his native Isle of Man, designed in a manner influenced by the numerous intricate, interlaced petroglyphs of the remaining Celtic monuments found on the island. Celtic design was, after all, the chosen subject of his final school examination. The mirror shown here and other Knox work, was produced by Arthur Lansenby Liberty's eponymous firm in London from 1898. Some of Liberty's output was not true to the lofty aspirations of the British Arts and Crafts Movements whose advocates espoused all-hand production. The Arts and Crafts proponents were reacting against poorly designed, machine-produced goods that had been encouraged by the Industrial Revolution. Nevertheless, Liberty frequently had his craftspeople stamp out or die-cast objects which were then finished by hand to make them appear more custom made. But, since Liberty did not allow individual designers to sign their pieces, no one today knows who did what. For Liberty, others, and himself, Knox's output of metalwork and fabrics was abundant. From the Isle of Man from 1900–04, he sent his designs to Liberty for execution, possibly including this mirror. After working for Liberty until about 1909 and living in London on and off, Knox traveled to the U.S. where he sought work with little success and so returned to Britain in 1913. He finally resettled on the Isle of Man where he painted and taught.

Date: *c.* 1902. Materials: Silver, enamel, glass mirror, wood. Manufacturer: Liberty and Company, London. Photograph: Virginia Museum of Fine Arts, Richmond, gift of Sydney and Frances Lewis. Photo: Katherine Wetzel. © Virginia Museum of Fine Arts.

Events of 1902:

Folkwang Museum, designed by Henry van de Velde, was opened, Hagen, Germany.

Skeleton of a Neanderthal man was reconstructed by Pierre Boule, France.

Premiers: Ritz Hotel, London; fan club, for English actor Lewis Waller; Crayola Crayons for children, U.S.; air conditioner, by Willis H. Carrier, U.S.; New York Stock Exchange building; commercial production of the light bulb with an osmium filament; commercial hearing aid, by Millar Hutchinson, U.S.; gas-powered lawnmower, U.S.; Pavlovian theory of reinforcement learning; Irish Channel crossing by air, by J.M. Bacon in a balloon; Teddy Bear, named after U.S. President Theodore Roosevelt.

Oskar Messter filmed *Salomé*, Germany.

Music: First of Elgar's "Pomp and Circumstance" marches (to 1930), U.K., and opera singer Enrico Caruso's first phonograph recording.

Literature: Conan Doyle's *The Hound of the Baskervilles*, Gide's *L'immoraliste*, D'Annunzio's *Francesca da Rimini*, Chekhov's play "Three Sisters," Potter's *Peter Rabbit*, Maeterlinck's verse drama "Monna Vanna."

Auto innovations included the drum brake, by Louis Renault, France; spark plug, by Robert Bosch, Germany; disk brake, by Frederick Lancaster, U.K.; electric motor starter, by Camille Jénatzy, France.

First practical airship, *Le Jaune*, by the Lebaudy brothers, France.

First-known set of laws, the code of Hammurabi, was discovered by a French expeditionary team, Middle East.

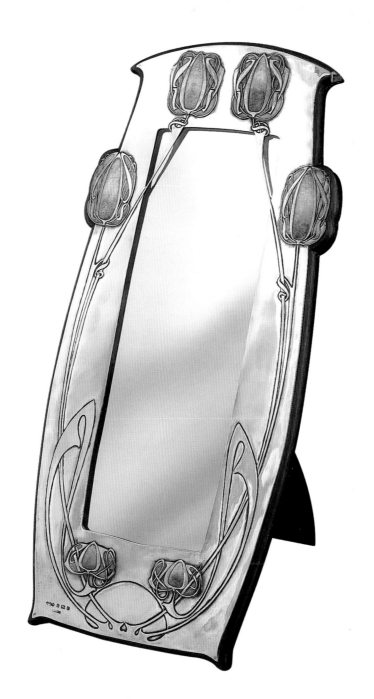

With his wife, her sister, and his brother-in-law, Charles Rennie Mackintosh (1868–1928) became active in architecture and design in his native Scotland. They formed "The Four" group while at the Glasgow School of Art. Unappreciated in his own country during his lifetime but adored by the Viennese at the time of the Vienna Secession and by the Germans, Mackintosh became a prolific designer of furniture, metalwork, textiles, glasswork, and graphics, which drew in part on Celtic ornamentation and the Japanese aesthetic. His Glasgow School of Art (completed in 1909) and interiors for hard-to-find clients are now considered brilliant examples that forecasted the stark Modernism that was to come. Mackintosh's architecture for publisher Walter W. Blackie's Hill House in Helensburg, Scotland, was devoid of naturalism, rather based on a functional, mathematical precision. Aspects of the imposing, rigidly geometrical, semi-functional chair shown here for Blackie's bedroom reveals both forward and regressive affinities in Mackintosh's work: the square-grid pattern employed by Mackintosh's Viennese contemporaries, the ladderback of traditional British seats, and the exaggerated heights of the chairs of the 1890s by Frank Lloyd Wright in America.

Date: 1903. Materials: Original in ebonized oak, fabric; reproduction shown here in black painted wood, fabric. Manufacturer: Alexander Martin, or other local joiner, Scotland; Cassina S.p.A., Meda (MI), Italy. Photograph: Courtesy Cassina S.p.A.

Events of 1903:

Wiener Werkstätte design workshop was opened by Vienna Secession members.

Premiers: The successful flight of a heavier-than-air machine, by the Wright brothers; Tour de France bicycle race; Hispano Suiza, first car in Spain; Harley-Davidson Motorcycles, U.S.; Royal Naval College, England; motor taxis, London; FAD (Fomento des las Artes Decoratativas), Barcelona; Metropolitan Opera House, New York; auto license plate, Massachusetts, U.S.; Sanka decaffeinated coffee; complete opera recording (Leoncavallo's "Pagliacci"); Ford and Buick auto companies; Prix Goncourt literary prize, France; Steuben Glass Works, U.S.; safety glass, accidentally developed by chemist Édouard Benedictus, France; experimental photocopying machine, by J.C. Beidler; electric locomotive, by Siemens, Germany; artificial silk (viscose), by Stearn and Topham; universities of Liverpool and Manchester, England.

Construction of the Metropolitan Cathedral, designed by G.G. Scott, was begun, Liverpool, England.

Autos reached speeds of up to 32 km/h., and the North American continent was crossed in 65 days.

Konstantin Tsiolkovsky proposed liquefied oxygen for space travel, Russia.

The Great Train Robbery became the longest film to date (12 minutes), U.S.

Literature: Shaw's play "Man and Superman," London's novel *The Call of the Wild*, Strindberg's historical play "Queen Christina."

Music: Delius's "Sea Drift," Manén's "Giovanna di Napoli," Elgar's oratorio "The Apostles," first performance of Puccini's "La Madame Butterfly," first all-African-American musical, Walker and Williams's "Dahomey," on Broadway, New York.

The youngest son of a rich coffee-plantation owner in Brazil, Alberto Santos-Dumont (1873–1932) lived an early isolated life as an enthusiastic fan of French science-fiction writer Jules Verne. At age 18, after inheriting the family fortune, Santos settled in Paris with dreams of making flight by humans possible. And indeed for a while, from 1906, everyone thought that this peculiar, diminutive man's Japanese-silk-and-bamboo craft had become the first to mechanically send a person aloft. Achieving some renown for the feat, he was even received by Teddy Roosevelt at the White House in Washington. However, though not properly witnessed, the American Wright brothers had two years earlier made the first mechanical flight. Nevertheless, Santos, who was to eventually commit suicide, made an indirect contribution of a different kind. Since Santos could not use a pocket watch while navigating his airplane, his friend the Parisian jeweler, Louis-Joseph Cartier (1875–1942), solved the problem and thus created the first commercially produced wristwatch. Watches have been worn on the wrist since Elizabethan times, but the idea never took hold until Cartier's appeared. Square, unlike all previous watches, with Machine Age characteristics of exposed screws and other elements, the "Santos" watch originally featured a second hand, gold case, leather strap, gold buckle, and sapphire winding crown.

Date: 1904. Materials: gold, clear glass, painted enamel, leather (reproduction shown here of 1990 of the original model of 1904). Manufacturer: Cartier, Paris, France. Photograph: Courtesy Cartier.

Events of 1904:

First garden city was built, England.

Autochromes (color films), for motion pictures, were invented by the Lumière brothers, France.

Using silk bags to send tea samples, importer Thomas Sullivan inadvertantly created the first tea bag when his customers mistakenly soaked the bags directly in boiling water.

Premiers: The flat disk photograph, improving Edison's cylinder, by German-American inventor Emile Berliner, adopted by the record industry as a standard; practical photoelectric cell, devised by Johann Phillip Ludwig Elster, Germany; stainless steel developed by Léon Guillet, France; offset printing method by W. Rubel; 10-hour work week, France; Georg Jensen, Copenhagen.

Work began on the Panama Canal, and W.C. Gorgas eradicated yellow fever which had broken out there.

Rolls-Royce was founded, Crewe, England.

Broadway line, the first subway in New York City, opened.

First female professor at the Sorbonne (1906), Marie Curie published Recherches sur les substances radioactives, France.

First photographs were transmitted telegraphically, by Arthur Korn from Munich to Nuremberg, Germany.

Louisiana Purchase Exposition and the third Olympics (and the first in the U.S.), St. Louis, Missouri, U.S.

Literature: Barrie's Peter Pan, Chekhov's play "The Cherry Orchard," Hesse's Peter Camenzind.

Church and state were separated, France.

Films: Le barbier de Seville and Le damation de Faust, France.

Helen Keller, the blind mute, graduated from Radcliffe College, Boston.

American architect and designer Frank Lloyd Wright (1867–1959) was a man of vision and had great imagination. So different are the oak barrel-back armchairs of the Darwin D. Martin House and the steel furniture of the Larkin Building, both in Buffalo, it seems far-fetched that they were designed by the same person at the same time. The Larkin Building, considered one of Wright's most important, was among the first to use air conditioning, plate glass, and wall-hung water closets. All of the clients for whom Wright designed houses in Buffalo were associated with the Larkin firm, a mail-order business founded by Arts and Crafts exponent Elbert Hubbard and his brother-in-law John Larkin. A tall brick structure, the Larkin administration building was one of a group of factories adjacent to railroad tracks in a grim section of the city. Wright designed the building to be light and beautiful inside but isolated from the outside. Both wood and metal furniture reflected the solid geometry of the building. Wright's plank-back chair, the forerunner of the Rietveld "Red Blue Chair" of 1918 (see p. 48), was used throughout including the cafeteria. Some of the desks (shown here) incorporated integral swing-out seats which permitted easy cleaning beneath. Separate, rolling-metal armchairs were used in the executive suite. Metal office furniture had been rarely used before this time, and Wright's metal office chairs and desks were the first that did not imitate wood.

Date: *c.* 1904–06. Materials: Painted steel, leather. Manufacturer: Van Dorn Iron Works Company, Cleveland, OH, U.S. Photograph: Virginia Museum of Fine Arts, Richmond, gift of Sydney and Frances Lewis. Photo: Ron Jennings. © Virginia Museum of Fine Arts.

Events of 1905:

New products included an early pressure cooker; "hot point" electric iron, by Richardson (later providing the name for the Hotpoint firm), and the "Westport" chair (along the lines of F.L. Wright and Rietveld plank chairs) by Thomas Lee, U.S.

Premiers: Rayon; patent for safety glass; Royal Typewriter Co.; L.C. Smith & Brothers typewriter; corporate identity program, for any firm, by Peter Behrens for A.E.G., Germany; plywood as laminated fir-wood sheets, by the Portland Manufacturing Company, U.S.; airplane factory by Voisin, Archdeacon, and Blériot, Billancourt, near Paris; intelligence test, by Binet, Henri, and Simon, France; chemical fire extinguisher; Austin Motor Company, England; motor buses, London; Piccadilly and Bakerloo underground (subway) lines, London; neon-light signage, by Georges Claude, Paris; direct blood transfusion, by George Washington Crile, U.S.; Novocaine, a local anesthetic, by Albert Einhorn; artificial human joints, by J.B. Murphy; discovery of the cause of syphilis, by Schaudinn and Hoffmann, Germany.

"291" gallery was opened by photographer Alfred Stieglitz, New York, and introduced Americans to modern art.

Albert Einstein received a doctoral degree and also this year submitted his first paper on the special theory of relativity, Switzerland.

First German U-boat was launched.

"Hormone" was coined by Ernest Starling in his paper "On the chemical correlation of the functions of the body."

3,000-carat Cullinan diamond, largest to date, was found, South Africa.

Desk and stool by Otto Wagner

The early work of Austrian architect, designer, and city planner Otto Wagner (1841–1918) was in the overwrought 19th-century style popular in Vienna at the time. But in 1895, the year after his becoming the leading professor of the Akademie der bildenden Künste in Vienna, Wagner turned to an early form of Modernism and became a major figure of the Vienna Secession movement. He had a great influence on European architecture through his own work as well as his pupils', who included Hoffmann and Olbrich (see pp. 14, 28). His most important contribution was the steel-and-glass-roofed Österreichische Postsparkasse (Imperial Austrian Postal Savings Bank) of 1904–06 and 1910–12 in Vienna. For its interior furnishings, Wagner combined traditional and modern materials—wood and aluminum—and made no effort to camouflage construction of the side chairs, armchairs, stools, file cabinets, and desks. Stools were installed throughout, but desks were placed in offices only. Aluminum strips on the arms and caps of the feet of the chairs for the boardroom served both to decorate and resist wear. Two different respected Viennese firms, Thonet and Kohn, received the manufacturing commission. The furniture, which signaled the first use of aluminum as a furniture element in Austria, was also sold for use by others and is in production today, somewhat altered.

Date: 1904–06. Materials: Brown-stained beechwood, aluminum. Manufacturers: Gebrüder Thonet (shown here) and Jacob und Josef Kohn, both Vienna, Austria. Photograph: Courtesy Die Neue Sammlung, Munich, Germany, inv. nos. 133/95, 134/95. Photo: Angela Bröhan, Munich.

Events of 1906:

Populations in millions reached 4.5 in London, 4 in New York City, 2.7 in Paris, 2 in Berlin, 1.9 in Tokyo, and 1.3 in Vienna.

Premiers: Thermos bottle, by Sir James Dewar, England; air conditioning, in F.L. Wright's Larkin Building, Buffalo; diesel locomotive engine, Germany; wall-hung version of telephone with exposed bells and dials (1891) for self-dialing, by Strowger, Chicago; French autobus, Paris; facsimile (fax or télécopier) machine, by Arthur Korn; Fuller Brush Company, selling directly to customers, U.S.; functional calculus, by Maurice Fréchet, France; medical term "allergy," coined by Clemens von Pirquet; Grand Prix motorcar race, France.

Famous test for syphilis was developed by bacteriologist August von Wasserman, Germany.

Marie Curie became the first female professor at the Sorbonne, assuming the position of her late husband, Paris.

Nazimova made her U.S. debut in the play "Hedda Gabler," New York City.

Ruth St. Denis introduced modern dance, U.S.

George M. Cohen produced *Forty-five Minutes from Broadway*, New York City.

Literature: Claudel's *Partage de midi*, Gjellerup's *The Pilgrim Kamanita*, O. Henry's *The Four Million*, Hauptmann's fairy-tale play "Und Pippa tanzt," Pinero's "His House in Order."

"Victrola" with a large sound horn was introduced by the Victor Talking Machine Company, U.S.

Mysterious explosion is unexplained until much later, near Tunguska, Siberia.

San Francisco earthquake killed about 700 people.

Like American architect Frank Lloyd Wright and others in Europe, Charles Sumner Greene (1868–1957) and Henry Mather Greene (1870–1954) built houses that integrated the architecture, landscape, and interiors. The Greene brothers' exquisitely refined aesthetic owed much to Japanese and Chinese forms and the highest aspirations of hand-hewning espoused by British Arts and Crafts exponents. Their interior woodwork, furniture, and fittings were oiled and buffed to a warm glow. Frequently specified in tropical woods, such as mahogany, the furniture featured exposed dowels and pegs in contrasting, exotic woods, like ebony. The lines and materials of machine-made furniture produced by other American Arts and Crafts workshops appear crude in comparison. Yet, like Wright, the Greenes installed Stickley furniture in their residences (see p. 40), until 1904, when they began concentrating on their own designs. The living-room-corner chandelier shown here in the same mahogany as the other furniture and furnishing in Greenes' R.R. Blacker House in Pasadena, exemplifies their manipulation of Oriental themes, fine woods, and other materials, including silver inlay and stained glass. With the best of their California "bungalows" (actually very large houses) of the 1907–09 period far behind, the brothers dissolved their partnership in 1922 and went their separate ways.

Date: 1907. Materials: Mahogany, silver inlay, glass, metal. Manufacturer: Peter Hall Manufacturing Company, Pasadena, CA, U.S. (frame); Emile Lang (stained glass). Photograph: Courtesy Randell L. Makinson, REM Associates, Pasadena, CA, U.S.

Events of 1907:

Premiers: The patent of Bakelite, first industrial plastic, by Dr. Leo Hendrik Baekeland, U.S.; comic strip "Mr. Mutt" (later "Mutt and Jeff"), U.S.; Deutsche Werkbund (German work union), Munich; Dryad furniture company, by Harry Peach, England; Converse Rubber Co., by Marquis M. Converse, U.S.; Curtiss motorcycle, by aviation pioneer Glenn Curtiss, U.S.; flight of the gyroplane, forerunner of the helicopter, by Gréguet-Richet, France; vertical take-off of a helicopter, by bicycle dealer Paul Cornu, France; Library wing of the Glasgow School of Art (to 1909), by Charles Rennie Mackintosh, Scotland; paint spray gun; Boy Scout movement, by Robert Baden-Powell, U.K.; ocean liners *Lusitania* and *Mauretania*, U.K.; modern zoo, by Carl Hagenbeck, Hamburg, Germany.

Film: Titles replaced commentators and, independently, slow-motion effect in motion pictures was invented by August Musger. *Skating*, with Max Linder was filmed, Germany.

Literature: Conrad's *The Secret Agent*, Gorki's *Mother*, Rolland's *The Life of Beethoven*.

Art: Picasso's *Les demoiselles d'Avignon*, birth of Cubism; Derain's *Blackfriars Bridge, London*; Rousseau's *The Snake Charmer*; Chagall's *Peasant Women*; Munch's *Portrait of Walter Rathenau*.

Basis for the study of modern linguistics was codified by Ferdinand de Saussure, France.

Siam (today Thailand) became independent by an agreement of the U.K. and France, and New Zealand became a dominion of the U.K.

Emperor of Korea abdicated, and Japan became the protectorate of Korea.

Like Wright, Wagner, the Greenes, and others (see pp. 22, 24, 26), Josef Hoffmann (1870–1956) approached a building as a total work of art, outside and inside. In order to fulfill this aspiration, Hoffmann and others in 1903 founded the Wiener Werkstätte, an art-craft workshop-cooperative that became as much a movement as an enterprise to make money. Due to the reproductions readily available today, Hoffmann may be best known for his table and chair for the Cabaret Fliedermaus in Vienna. His architecture included the impressive Palais Stocklet (1905–24) and the Pukersdorf Sanatorium (1904–05). It was for the sanatorium that he designed the "Sitzmaschine" (machine for sitting). The armchair was first shown in a model country house at the Kunstschau of 1908 in Vienna. The exhibition served to display furniture made by the Kohn firm. The "Sitzmaschine" may have been based on the well-known Morris chair of the 1860s by William Morris's firm in England. Like the Morris model, Hoffmann's chair with his characteristic use of wooden balls also has an adjustable back rest. Yet, by utilizing wood that was bent by a technique that had become highly developed in Austria, Hoffmann transcended the traditional English easy chair and produced a geometric *tour-de-force* sympathetic with the art reform of Vienna at the turn of the century.

Date: 1908. Materials: Stained beechwood, metal. Manufacturer: Jacob und Josef Kohn, Vienna. Photograph: Courtesy Die Neue Sammlung, Munich, Germany, inv. no. 894/88 Photo: Bayerisches Nationalmuseum.

Events of 1908:

Swiss Officers' Knife was first sold to the Swiss Army.

Premiers: Ford "Model T" auto; General Motors, U.S.; Olivetti, Italy; steel-and-glass building, A.E.G. turbine factory by Peter Behrens, Berlin; 9mm automatic pistol, by G. Luger, Germany; method for cellophane production, by Swiss chemist Jacques Edwin Brandenberger, France; patent for portable vacuum cleaner, by James Murray Spangler and introduced by Hoover, U.S.; tractor with moving treads, by The Holt Company, California; black heavyweight boxing champion (Jack Johnson), U.S.

"Ornament und Verbrechen" (Ornament and Crime), the essay against excessive surface decoration, was published by Adolf Loos, Germany.

Dancer Isadora Duncan became popular, U.S.

French jeweler Pierre Cartier acquired the 45.5-carat Hope diamond from the Hope estate in London.

The Times newspaper was bought by Lord Northcliffe, London.

Film: Ambrosio's *The Last Days of Pompeii* and the one-reel *Dr. Jekyll and Mr. Hyde*, the first horror film.

Literature: Chesterton's *The Man Who Was Thursday*, Colette's *La retraite sentimentale*, Forster's *A Room with a View*, France's *L'île des pingouins*, Stein's *Three Lives*.

Matisse coined the art term "Cubism," Paris, and Jacob Epstein's *Figures* at the British Medical Association Building, London, caused general indignation.

Music: Bartók's "String Quartet No. 1," O. Straus's operetta "The Chocolate Soldier," Fall's operetta "The Girl in the Train."

Union of South Africa was established.

Electric kettle by Peter Behrens

The profound influence of Peter Behrens (1868–1940) on architecture and design has continued to today. In the year of 1910 alone, pioneering architects and theoreticians Le Corbusier, Ludwig Mies van der Rohe, and Walter Gropius all worked side-by-side in the Behrens office in Berlin. Behrens's career spanned the Jugendstil (or German Art Nouveau), Arts and Crafts movement, rectilinear Modernism, and neoclassicism. He was teaching in Düsseldorf in 1907 when he was invited by Walter Rathenau, managing director of A.E.G., to design advertising and packaging for the huge electrical combine in Berlin. Eventually he worked on all facets of the firm's industrial design, including architecture, the corporate identity, graphics, clocks, electric fans, kettles, and lamps. This was the first time any firm in the world had hired an artist to advise on all its industrial design, or to assume the position of artistic director. The A.E.G. kettle shown here, the first with a submersible heating element, was specified by Behrens to be assembled from a choice of 3 bodies, 2 lids, 2 handles, and 2 bases—24 possible versions. The hammered effect and cane-covered handle suggested handcrafting. The A.E.G. foray into modern industrial design was an isolated episode in Behrens's career, and, by 1907, he had turned to the neoclassical.

Date: 1908–09. Materials: Brass, cane, Bakelite. Manufacturer: A.E.G. (Allgemeine Elektricitäts-Gesellschaft), Berlin, Germany. Photograph: Courtesy Die Neue Sammlung, Munich, Germany, inv. no. 44/74. Photo: Angela Bröhan, Munich.

Events of 1909:

Frank Lloyd Wright built the Robie House, Chicago.

Premiers: The manufacture of Bakelite (patented 1907), by Leo Baekeland, U.S.; electric toaster, by General Electric, U.S.; Alfa Romeo autoworks, Italy; commercial aeronautics firm, established by Count von Zeppelin who transported 100,000 people from 1909–14, Germany; crossing of the English Channel by air, by French pilot Louis Blériot; airplane flight of 100 miles, by British pilot Henri Farman; hydrofoil ship, by Enrico Forlanini; permanent hair waves for women, London; rayon stockings for women, Germany; Selfridge's Department Store, by American H.G. Selfridge, London; Kibbutz, in Palestine.

Original manifesto of Futurism by Filippo Marinetti, the Italian writer-theoretician, was published in *Le Figaro*, Paris.

Ballets Russes de Monte Carlo was installed in Paris by Sergei Diaghilev.

Women were admitted to German universities.

Anglo-Persian Oil Company was founded.

Literature: Maeterlinck's fairy-tale *L'oiseau blue*, Apollinaire's *L'enchanteur pourrissant*, Mann's novel *Königliche Hoheit*, Molnár's play "Liliom."

Music: Delius's "A Mass of Life," Strauss's opera "Elektra," Mahler's "Symphony No. 9," Williams's "Fantasia on a Theme of Tallis," Schönberg's "Three Piano Pieces" (Op. 11).

Sigmund Freud lectured on psychoanalysis in the U.S.

That typhus fever was transmitted by a body louse was discovered by Charles-Jules-Henri Nicolle, France.

Caran d'Ache and Frederick Remington died.

"M1" typewriter by Camillo Olivetti

Before Camillo Olivetti (1868–1943) founded his eponymous typewriter company in 1908 in Italy, he had made a trip to the U.S. to observe manufacturing techniques there and may have studied the production of typewriters first hand. Distinct from the lively decorative arts in Europe at the turn of the century, ideas coming from America had a greater impact on Italian industrial design. By 1867, the first true, workable version of a typewriter had been built by American inventor Christopher Sholes. Rights to his faster-typing version of 1873 were acquired by Remington, the gunsmith, who introduced a crude commercial model the next year. However, the Underwood model of 1895 that allowed a writer to see words as they were being typed may have been the model on which Olivetti based his "M1." Even though Olivetti's machine was not technologically advanced, its good looks encouraged the thrust of Italian industry toward a sophisticated aesthetic. In 1911, Olivetti showed it with great pride at the International Exposition of Industry and Labor in Turin, whose title reflected the country's incipient focus on the value of industrial production. Frequently assigned a date of 1911, the development of the "M1" probably occurred between 1908 to 1910 and may have been completed just before 1911 when, from March, it was on display at the exposition in Turin.

Date: 1908–10. Material: Painted steel. Manufacturer: Ing. C. Olivetti & C., Ivrea, Italy. Photograph: Courtesy Archivio Storico Olivetti.

Events of 1910:

Exposition universelle et internationale opened, Brussels.

South American tango dance gained great popularity in Europe and the U.S.

122,000 telephones were in use, U.K.

The week-end became popular, U.S.

Premiers: The trench coat (Tielocken coat), by Burberry, England; food mixer, by Hamilton Beach; gas-propelled combine harvester, U.S.; "turtle," precursor of the modern robot, by British-American neurologist William Grey Walter; electric washing machine; machine-shop forerunner of Black & Decker, U.S.; tubular steel, in the Fokker "Spider Mark 1" airplane, Germany; double-decker bus, London; road map, by Michelin, France.

Ausgeführte Bauten und Entwürfe von Frank Lloyd Wright (The Buildings and Projects of Frank Lloyd Wright), the two-volume portfolio that was to have a powerful impact on European architecture, was published by Ernst Wasmuth, Berlin.

Eugene Ely became the first person to take off in an airplane from the deck of a ship, paving the way for the aircraft carrier.

Literature: Claudel's *Cinq grandes odes,* Forster's *Howard's End*, Péguy's play "Le mystère de la charité de Jeanne d'Arc," May's novel for boys *Winnetou,* Wedekind's play "Schloss Wetterstein," Michaelis's novel *The Dangerous Age.*

Film: *A Child of the Ghetto, Messaline, Lucrezia Borgia, Hamlet, Peter the Great.*

France executed Mata Hari for espionage, but as a cover up for governmental ineptitude.

In Prague, Josef Chochol (1880–1956) and other architects were the first generation of pupils of Jan Kotera who established Czech Cubism as a program to pursue the manipulation of plastic masses. With lofty aspirations, they developed a new language in design and architecture that would integrate their spiritual and psychological theories into a rhythmical play of triangles, crystals, and other "privileged forms." And, indeed, the emotionally based furniture and architecture drastically broke with tradition. Chochol, who studied in Prague and under Otto Wagner (see p. 24) in Vienna, was the architect of a number of houses in Prague and the now-demolished Barikádniku Bridge. The chair shown here was part of a suite, including tables, produced for the English Circle salon-exhibition at the Municipal Hall in Prague. Chochol's chair, more like an ancient throne fitted with wheels and covered with a monochromatic fabric, validated his mantra of 1913: "We do not wish to disrupt that rare, clean, smooth effect of a modern creation—austere as well as fantastic—by unbearable impediments of multitudes of indifferent little ornaments and details." He left Cubism behind in 1914 and, by the 1920s, had become influenced by the simplicity of and theories behind Russian Constructivism.

Date: 1911. Materials: Black-stained oak, fabric upholstery (replaced 1973), metal. Photograph: Courtesy Umelecko-prumyslové Muzeum v Praze (Museum of Decorative Arts), Prague, Czech Republic. Photo: Miloslav Sebek.

Events of 1911:

45% of the French population lived in cities.

Temperature in London reached 100° F.

Premiers: Oreos, later to become the world's most popular cookie, U.S; "6 CV" auto, by Ettore Bugatti and built by Peugeot, France; practical self-starter for autos, by Charles Franklin Kettering, U.S.; airplane with an enclosed passenger cabin (the *Berline*), by Louis Blériot, France; Eiffel Tower radio transmitter, Paris; Der Blaue Reiter avant-garde artists' group, Munich; gyrocompass, by Elmer Ambrose Sperry, U.S.; stainless steel as being resistant to corrosion, by P. Monnartz, Germany; intercontinental flight, by Calbraith P. Rogers, New York to California; Crisco hydrogenated vegetable cooking oil, U.S.; seaplane.

Fagus factory, with an expansive clear glass-enclosed staircase, by Walter Gropius and Adolf Meyer, Alfeld and der Leine, Germany.

Mona Lisa painting by Leonardo was stolen from the Louvre, Paris, by an Italian nationalist (recovered 1913).

Jacob Epstein designed the tomb of Oscar Wilde, Paris.

Film: *Anna Karenina*, *Spartacus*, *Pinocchio*, *Nick Carter*, *Enoch Arden*, *The Abyss* with Asta Nielsen.

Literature: Beerbohm's *Zuleika Dobson*, Dreiser's *Jennie Gerhardt*, Lawrence's *The White Peacock*, Wharton's *Ethan Frome*, Chesterton's *The Innocence of Father Brown*, Taussig's *Principles of Economics*.

Music: R. Strauss's opera "De Rosenkavalier," Ravel's "L'heure espagnole," Stravinsky's "Petrouchka," Berlin's musical "Alexander's Ragtime Band."

Revolution in Central China: Chinese Republic was proclaimed; pigtails were abolished; and the calendar was reformed.

Sometime between 1911 and 1913, engineer and amateur photographer Oskar Barnack (1879–1936) of Wetzlar, Germany, designed a small metal camera that used standard perforated movie film. After several permutations and a delay due to World War I, the camera was put into production by Barnack's employer, E. Leitz, a manufacturer of high-quality optical instruments. The first version, the "Pilot" of 1923–24, was not a prototype but rather manufactured in a limited series of 30 and sold to customers. The "Pilot" featured a focusing knob; a stationary, collapsible lens; and controls located on the top of the body. The film could be rewound, and instantaneous or time-elapse exposures could be made. The model that followed was the "Leica A" of 1925; it had a slow-shutter speed capability which was absent from the "Model B," introduced the next year. The "Leica" had established the new standard for picture taking and altered camera design and engineering forever. As the first compact device to make professional-quality images, it quickly became the camera of choice by photojournalists worldwide. The name "Leica" was derived from "Lei" in "Leitz" and "ca" in "camera."

Date: 1911–13. (The "Pilot" model here is no. 109 of nos. 101–130.) Materials: Leather-covered metal, black-painted brass fittings, removable base plate, chemically blackened lens mount. Manufacturer: E. Leitz, Wetzlar, Germany. Photograph: Courtesy George Eastman House, Rochester, NY, U.S.

Events of 1912:

Titanic ocean liner, considered unsinkable, hit an iceberg and sank in the North Atlantic Ocean; 1,500 people died.

Monotype Corporation produced "Imprint," the first original typeface for mechanical composition.

Prizes were added to Cracker Jacks, the candy-coated popcorn, U.S.

Premiers: Parachuting from an airplane; disposable Dixie Cup, by Charles M. Gage, to avoid disease risk from public drinking; electric headlamp on an auto, on the Cadillac; heating pad, by Dr. Sidney Russell, U.S.; laboratory of aerodynamics, by Gustave Eiffel, Auteil, France; cellophane process, by Edwin Brandenberger; self-service grocery store, California; exhibition of contemporary design in the U.S., by John Cotton Dana, Newark (NJ) Museum, U.S.; Universal Pictures movie studio, Hollywood; *Pravda* newspaper, by the Bolsheviks, Moscow; L.L. Bean hunting-supplies store, Freeport, Maine, U.S.; New Delhi, to replace Calcutta as the Indian capital; medical human-knee hammer for leg responses.

Art: Chagall's *The Cattle Dealer*, Picasso's *The Violin*, Frampton's *Peter Pan* in Kensington Gardens, London.

Literature: Hauptmann's *Atlantis*, Claudel's *L'annonce fait à Marie*, Maugham's social drama *The Land of Promise*, Jung's *Psychology of the Unconscious*.

Music: Ravel's ballet "Daphnis et Chloë," Schönberg's song cycle "Pierrot Lunaire," Frinl's operetta "The Firefly." Stokowski became conductor of the Philadelphia Symphony Orchestra, U.S.

English explorer Robert Falcon Scott reached the South Pole; however, Amundsen had arrived a month before.

The word "vitamin" for a class of "accessory food factors" was coined by Casimir Funk, U.S.

Nikolai Anichkov declared that cholesterol was responsible for coronary artery disease.

Armchair by Francis Jourdain

Francis Jourdain (1876–1958) was a member of a distinguished family of architects in France that included his father Frantz Jourdain whose masterpiece was the Samaritaine department store of 1905–07 in Paris. Francis studied painting and showed his canvases alongside those of Cézanne, Matisse, de Toulouse-Lautrec, and Kandinsky; wrote about art; was a novelist; and eventually published his memoirs. He became strongly influenced by the treatise of Viennese architect Adolf Loos, "Ornament and Crime" of 1908, which railed against excessive surface ornamentation. Taking Loos to heart, Jourdain produced designs for simple, economical furniture that could be mass manufactured. Quite active, he also designed interiors and furnishings for a number of clients. The chair shown here—part of a suite in Jourdain's own apartment in Paris—represents his penchant for flexible, purely utilitarian objects and furniture with multiple uses. In Paris, it was important for a designer to participate in annual design events, known as Salons. Thus, in 1913, Jourdain installed the furnishings of his own living room at one of the venues, the Salon d'Automne. At this time when Art Nouveau was in decline, Jourdain's work was jarring in contrast to much of the overwrought work of others and received negative criticism.

Date: 1913. Materials: Mahogany, split cane. Manufacturer: Ateliers Modernes, Esbly, France. Photograph: The Metropolitan Museum of Art, Purchase, Lita Annenberg Hazen Charitable Trust Gift, 1985 (1985.54.1).

Events of 1913:

Woolworth Building became the tallest in the world, and Grand Central Terminal was completed, New York.

The term "industrial design" was first publicly used by Edward B. Moore, the U.S. Commissioner of Patents who favored altering existing relations to protect intellectual property.

Premiers: Omega workshops, England; *De Stijl* journal (see p. 48), Netherlands; sale of the home refrigerator, Chicago; Citroën, France; crossword puzzle, by English-born journalist Arthur Wynne, U.S.; Erector set child's toy, by A.C. Gilbert, U.S.; radio transmitter with vacuum tubes, by Alexander Meissner, U.S.; vacuum triodes to amplify weak telephone signals; industrially produced cellulose acetate; wire wheels on standard-production autos; mammography, by A. Salomen, Germany; airplane flight to last an hour, by the Wright brothers, U.S.; artificial kidney, by John Jacob Abel, U.S.; ozone layer, discovered by Charles Fabry, France; female magistrate, England.

Literature: Marcel Proust's *Du côté de chez Swann* (first part of *À la recherche du temps perdu)* and Sigmund Freud's *Totem and Taboo.*

Music: De Falla's opera "Vida Breve," Stravinsky's ballet "Le sacre du printemps," Debussy's ballet "Jeux," music hall comedian Jack Judge's song "Tipperary," Victor Herbert's operetta "Sweethearts."

First films of Charlie Chaplin and of Paramount Studio.

New York Armory Show introduced Cubism and modern art to America.

The fox trot dance became popular.

John Pierpont Morgan and Montgomery Ward died.

American proponents of the British Arts and Crafts movement drew more from its superficial and aesthetic aspects than from the British fervor and deep commitment to the utopian ideals fomented in a reaction against the shoddy goods that the Industrial Revolution had made possible. American Arts and Crafts examples were based more on a mood than a definite style. The finest expressions were produced by the Greene brothers (see p. 26) in California. Yet, the success of the movement in the U.S. was due in part to the inexpensive wares, particularly those of Gustav Stickley (1858–1942) and his four brothers, available to middle-class Americans. Gustav Stickley's parents, who anglicized their German name Stoeckel, settled in Pennsylvania where Gustav first worked in his uncle's chair factory. From the 1880s, Gustav worked in his own furniture firm in Binghamton, New York, soon began designing in an Arts and Crafts style, and was in competition with his brothers. Gustav went on to become the greatest single influence on the American Arts and Crafts movement. His expansive empire included the publication of a magazine, sales of franchises, and offices in New York City. His table clock shown here with exposed pegs and joinery, insinuating total hand production, was primarily produced with machinery. The use of the machine was far less contentious in the U.S. than in Britain, though for some Americans it meant art in a folksy, if not highly superficial manner.

Date: *c.* 1912–16. Materials: Quarter-sawn American white oak, brass, other metals, Seth Thomas movement. Manufacturer: United Crafts Workshop, Morris Plains, NJ, U.S. Photograph: Virginia Museum of Fine Arts, Richmond, gift of Sydney and Frances Lewis. Photo: Ron Jennings. © Virginia Museum of Fine Arts.

Events of 1914:

Steel furniture for the battleship *Von Hindenburg* was designed by Walter Gropius and Adolf Meyer, Germany.

Premiers: Cellucotton, developed by Kimberly-Clark, later used to make Kotex menstruation napkins and Kleenex face tissue, U.S.; modern sewage plant, to treat sewage with bacteria, Manchester, England; opening of Panama Canal; patent for the brassiere; self-pulling static line of parachutes, by Tiny Broderick, U.S.; teletypewriter, by Edward Kleinschmidt, Germany; Brooks Sports (sports shoes); red and green traffic lights, Cleveland, Ohio, U.S.

"Ready Made" art concept was presented by Marcel Duchamp.

In the book *Stellar Movements and the Structure of the Universe*, Arthur Stanley Eddington suggested that spiral nebulae are galaxies, U.K.

Robert Goddard began experimenting with rocketry, U.S.

First successful heart surgery was performed by Dr. Alexis Carrel, on a dog, Massachusetts, U.S.

Film: *Making a Living* with Chaplin, U.S.; *The Golem*, Germany; *The Little Angel*, Denmark; Sennett's *Tillie's Punctured Romance*, first full-length comedy.

Literature: Joyce's *Dubliners*; Kilmer's poem "Trees"; Gide's *Les caves du Vatican*; Rice's "On Trial," the first drama to use the technique of the flashback; Paul Bourget's *Le démon de midi*; Burroughs's *Tarzan of the Apes*.

World War I was precipitated by the assassination of Archduke Ferdinand (heir to Austrian throne), Serbia.

Approximately 10,500,000 immigrants had entered the U.S. from southern and eastern Europe (1909–1914).

In 1886 in Atlanta, Georgia, American pharmacist John S. Pemberton invented the formula for Coca-Cola, a concoction of cola-nut extract, sugar, caffeine, and other ingredients. A year later, drugstore barman Willis Venables mixed the Coca-Cola syrup with carbonated water and sold a glassful for 5¢. Unable to raise the money to promote the drink properly, Pemberton sold his shares in 1890 to pharmacy owner Asa G. Candler who founded the Coca-Cola Company. The drink had been sold only at drugstore soda fountains before it was first bottled in 1894. The business prospered and already had a very large advertising budget by 1901. Some stories tell of the later, hobble-shirt Coca-Cola bottle having been first designed by machinist Earl Dean and that the handwritten trademark was created by Pemberton's bookkeeper. But general credit for the famous bottle design is today being assigned to Swedish engineer Alexander Samuelson of the Root Glass Company of Terre Haute, Indiana. His version of the nipped-waist shape was purportedly based on an illustration he had seen of a cocoa bean in *Encyclopedia Britannica*. The glass bottle, patented in 1915 and introduced in 1917, was made in clear, aqua, ice-blue, or green tints. It was exclusively green after Christmas day 1923, except during the war when it was blue due to a copper shortage.

Date: Patented 1915. Materials: Blue-tinted glass (example shown here 1915–1917). Manufacturer: Coca-Cola Company, Palaka, FL, U.S. Photograph: © Greg S. Krum, New York. Bottle courtesy Bill Porter, Washington, DC.

Events of 1915:

Panama-Pacific International Exposition opened, San Francisco, and featured spectacular illumination and commemorations of the accomplishment of Cortez, Pizarro, Columbus, and the Panama Canal (19,000,000 visitors).

Design and Industries Association, based on the Deutsche Werkbund's aspirations to remedy bad design in the marketplace, was established, London.

Premiers: Borosilicate glass with high-heat resistance, at the Corning Glass Works; gas-powered clothes washing machine, by Maytag, U.S.; patent for a bath scale, by M.J. Weber; tractor trailer, by blacksmith August Freuhauf, Detroit, U.S.; sonar, mainly for ships to detect icebergs, by scientist Paul Langevin, France; motorized taxis, London; general theory of relativity, describing gravity as a curved space, by Albert Einstein; farm tractor, by Ford; transcontinental telephone call, New York to San Francisco; transatlantic radio-telephone conversation, Arlington, Virginia, to Paris; largest railroad station in Europe, Leipzig; fighter airplane, by Hugo Junkers, Germany.

Marcel Duchamp, a leading exponent of the Dada movement, not to be fully formed until 1916, began producing the first Dada-style paintings, New York.

Literature: Buchan's *The Thirty-Nine Steps*, Claudel's *Corona*, Lawrence's *The Rainbow*, Maugham's *Of Human Bondage*, Hesse's *Knulp*.

Ivor Novello wrote the war song "Keep the Home Fires Burning."

The remains of Rouget de Lisle, composer of "La Marseillaise," were brought to the Invalides, Paris.

Wireless radio service was established between Japan and U.S.

An *ensemblier*, or decorator, Émile-Jacques Ruhlmann (1879–1933) took over his family's house-painting business in the first decade of the 20th century. He began showing his work in Paris in 1911. His early furnishings, like the example shown here, were not modern but rather influenced by French period styles of the 18th and 19th centuries and foretold what became Le Style 1925. Ruhlmann's exquisite creations were singularly his own, unequaled by his contemporaries' in both style and craftsmanship. The cabinet shown here combined rich, rare woods and other materials that produced a visual feast. The vase motif on the lacquered-rosewood door explodes in an exuberant bouquet of flowers painstakingly created by inlaying a variety of imported wood veneers and ivory. The ivory, readily available from French colonies in Africa, was both inset as flat planes flushed to surfaces and carved for the feet and leg knees. Ruhlmann's chairs, tables, wallpaper, lamps, and other furnishings were produced by artisans in his own workshop who took as long as a year to fulfill orders, frequently to customers' dismay. Working with his partner and a painting contractor, Pierre Laurent, from 1919, Ruhlmann's work by the mid-1920s had become far simpler in form but retained his predilection for precious materials and superb craftsmanship.

Date: 1916. Materials: Lacquered rosewood, rare woods, ivory. Manufacturer: Société Ruhlmann, Paris. Photograph: Virginia Museum of Fine Arts, Richmond. Gift of Sydney and Frances Lewis. Photo: Katherine Wetzel. © Virginia Museum of Fine Arts.

Events of 1916:

Disposable cups, "Health Kups "(1912), later named "Dixie Cups," were first produced by Hugh Moore, U.S.

Converse Rubber Co. produced the first "All-Star" canvas-and-rubber sports shoe, U.S.

Identity program for the Underground was developed by Frank Pick, with designs by Edward Johnston, London.

"Coco" Chanel established her fashion house, Paris.

Premiers: Refrigerated blood; guide dogs for the blind, Germany; car windshield wiper; lipstick in metal cartridges; radio tuning device; Lucky Strike cigarettes; daylight-savings time ("summertime"), U.K.

Developed in 1904 by French scientist Léon Guillet, stainless steel was patented by others.

Hangers to house dirigibles were designed by Eugéne Freyssinet and built at Orly, France.

Imperial Hotel, later to resist a strong earthquake, was built by Frank Lloyd Wright, Tokyo.

Dada, the nihilistic art cult, was initiated, Zürich.

Radio was transmitted across the Atlantic Ocean.

Film: D.W. Griffith's *Intolerance*; *The Pawn Shop* and *The Immigrant* with Charlie Chaplin; *Resurrection*; and *Homunkulus*, in 5 parts, Germany.

Literature: Ibáñez's *The Four Horsemen of the Apocalypse*, Joyce's *Portrait of the Artist as a Young Man*, Andreyev's *He, the One Who Gets Slapped*.

Spurred by the first New Orleans-style jazz recording, the style swept the U.S.

U.S. entered World War I on the allies' side.

The Royal Air Force recommended that Stonehenge be demolished because it was hazardous to low-flying aircraft, an idea rejected by then-owner Cecil Chubb, U.K.

Rasputin was murdered.

Elsa Gullberg (1886–1984), a textile designer active in efforts to improve the applied-arts industries of Sweden, arranged a number of artist-manufacturer collaborations including the relationship of painter and drawing teacher Edvin Ollers (1888–1959) and the Kosta glassworks. Ollers was hired to represent the factory at the Home Exhibition of 1917 at the Liljevalchs Konsthall in Stockholm. Kosta's efforts were a reaction to the nascent social consciousness developing in Scandinavia in the early 20th century that sought to provide attractive, inexpensive products to the public. The example shown here, unlike some others, was a one-of-a-kind piece. The vase was designed by Ollers but produced by blowers and other factory workers who flashed a thick layer of yellow glass onto the outside. Oller's vegetal motif was then traced onto the surface, painted over with an acid-resistant adhesive, and placed into baths of hydrochloric acid until the background of the motif was eaten away. Ollers stayed for only a short time at Kosta, originally a window-glass factory founded in 1742. Kosta is an acronym of the names of its founders, A. Koskull and Bogislaus Staël von Halstein. In 1946, Kosta merged with other glass factories, Boda and Åfors, and, in 1990, was further consolidated with Orrefors to form Orrefors Kosta Boda.

Date: 1917. Materials: Blown, overlay-etched glass.
Manufacturer: Kosta Glasbruk, Kosta, Sweden. Photograph:
Courtesy Nationalmuseum, Stockholm, Sweden, inv. no.
NMK 236/1967.

Events of 1917:

America declared war on Germany.

In Russia, February and October Revolutions overthrew the Czarist regime; Constructivism emerged; State Porcelain Factory, employing the artistry of Contructivist painters and designers was re-established from the Czar's Imperial Porcelain Factory, Petrograd.

Premiers: Popularly marketed sneakers (athletic footwear "Keds"), by U.S. Rubber, U.S.; steel-wool pot-cleansing pads infused with soap (S.O.S. pads), by Edwin W. Cox; windshield wipers for automobiles, U.S.; modern washing machine, by John Fisher, U.S.

Clarence Birdseye developed freezing as a method for preserving food, U.S.

Film: *The Little Princess* with Pickford and *Mater dolorose*. U.F.A. (Universum Film, Berlin) became the foremost production company.

Literature: Eliot's *Prufrock and Other Observations*, Feuchtwanger's *Jud Süss*, Valéry's *La jeune parque*, Barrie's *Dear Brutus*, Freud's *Introduction to Psychoanalysi*s.

Picasso designed the Surrealist stage set and costumes for Eric Satie's ballet "Parade," Paris.

Music: Respighi's four symphonic poems "Fontaine di Roma," Busoni's two one-act operas "Turandot" and "Harlequin," Sigmund Romberg's operetta "Maytime."

Sarah Bernhardt (age 72) began her last tour of the U.S.

World War I continued: bread was rationed, U.K.; air attacks began on England; first U.S. Army division arrived in France.

Gerrit Thomas Rietveld (1888–1964) became a member of De Stijl, a group of Dutch architects, artists, and writers lead by Theo Van Doesburg, who were active from 1917–1931. They attempted to marry, rather than separate, fine art and the applied arts; from the time he was in school, Rietveld had felt that painting was important in the development of a whole new style of architecture and design. With much in common with the geometrical planes of his canonical Schroeder house in Utrecht, Rietveld designed a black-and-white armchair in 1917—far less energetic, of course, than the final and highly uncomfortable expression shown here that was created a year later. The 17 different parts and pieces of the armchair give the illusion of merely kissing each other; yet, the 7 rails and 6 posts are held tightly to each other by dowels. Plain-painted high and low versions were also made in 1918, presumably for a man and a woman. Rietveld may have been inspired by the plank elements incorporated into the chairs of American architect Frank Lloyd Wright and Dutch architect H.P. Berlage. However, Rietveld's use of space and dynamic colors, which he began using more and more from 1922, is essentially absent in their work.

Dates: 1917 (unpainted or black), 1918 (painted red, blue, yellow, and black). Materials: Deal wood, plywood (original). Manufacturers: Gerald A. van de Groenekan, Utrecht (from 1918); Cassina, Meda (MI), Italy (shown here). Photograph: Courtesy Cassina.

Events of 1918:

Women were first allowed to vote, England.

U.S. Post Office burned copies of the *Little Review* in which James Joyce's *Ulysses* appeared in installments.

Premiers: 3-color traffic lights, New York; electric refrigerator, by General Motors; granulated laundry soap (Rinso), by Lever Brothers; Ragged Ann doll; Kotex menstruation napkins; air-mail postage; electric food beater; daylight-saving time, all U.S.

Film: Lubitsch's *The Eyes of the Mummy* (*Die Augen der Mummie Ma*) and *Carmen*; *Shoulder Arms* and *The Kid* with Charlie Chaplin.

Art: Mondrian's *Composition No. 2*, Modigliani's linear-style painting *Act*, Nash's *We Are Making a New World*, Matisse's *Odalisques*, and Joan Miró's first exhibited work.

Music: Berlin's musical "Yip, Yip Yaphank," and Stravinsky's "Histoire du soldat" (the soldier's story), Lausanne.

Despite daily bombing, the Opéra opened with Charles Gounod's "Faust," Paris.

Reacting to World War I, New York Philharmonic Society banned works by living German composers.

Excavations of Babylonia, and later Ur, were begun by British archaeologist Leonard Woolley.

Influenza pandemic killed 20- to 40,000,000 people worldwide, more than the combined total of the Black Death, World War I, and the future World War II.

Matsushita Industries was established, Japan.

World War I ended.

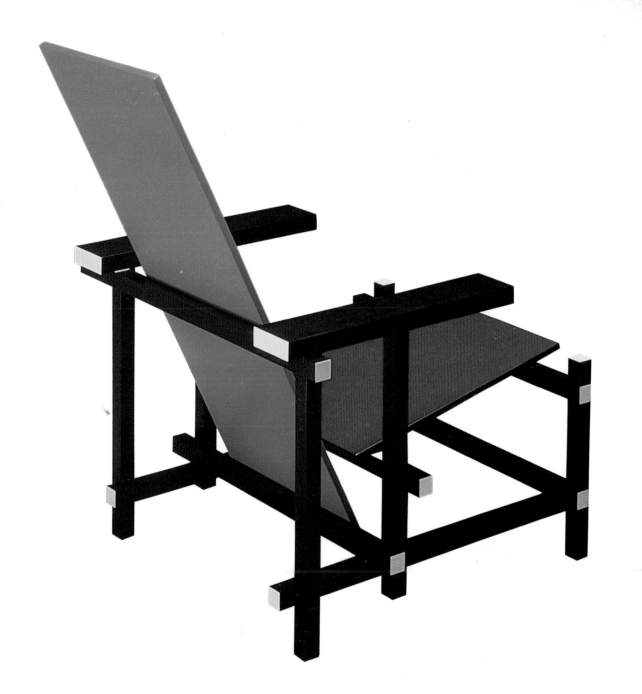

Jar by Gerhard Henning

Primarily known for his ceramic figural groups, Gerhard Henning (1880–1967) dressed his exotic people in bizarre, Oriental costumes. The decorative, not functional, jar shown here amalgamated a traditional 18th-century vocabulary with a simpler 20th-century voice. When Swedish and Danish decorative arts were being compared at the time of this piece, in the 1910s, Swedish critic Erik Wettergren felt that the work in Denmark was better because fine artists there worked in the applied-arts industries. Likewise, Henning, who was ironically Swedish, worked at the Danish porcelain factory Royal Copenhagen. At the end of World War I, Denmark began experiencing an incipient democracy and was becoming increasingly liberal. In this atmosphere, the roots of Danish studio pottery took hold. Even so, the commitment of a major industrial factory like Royal Copenhagen to support an experimental art studio was commendable. From the time of Henning's jar here to 1940, figural stoneware and Chinese and Japanese glazes were specialities at Royal Copenhagen. And it was there that the skilled potter Carl Haller worked with Henning and others including Jais Nielsen, Knud Kynd, and Jean-René Gauguin (the son of the painter). Examples of the collaboration were shown at the Paris exposition of 1925, the year Henning left Royal Copenhagen.

Date: *c.* 1919–20. Materials: Glazed porcelain. Manufacturer: Royal Copenhagen, Copenhagen, Denmark. Photograph: The Metropolitan Museum of Art, Purchase, Edward C. Moore, Jr., Gift, 1923 (23.114.5ab).

Events of 1919:

Two art schools became the Bauhaus art, architecture, and industrial-design school, Weimar, Germany.

Fiat factory began production, Lingotto, Italy.

Jean Patou opened his fashion house, Paris.

Premiers: The meeting of the League of Nations, Paris; non-stop air crossing of the Atlantic, by John Alcock and Arthur Brown, U.K.; commercial airline service, by Deutsche Luftreederie, Germany; pogo stick; gasoline tax, Oregon, U.S.; dial telephone, U.S.

Father of American rocketry, Robert Goddard, was ridiculed for suggesting that a small vehicle could be sent to the Moon.

Edwin Lutyens designed the Cenotaph, London.

Constructivist exhibition was held, Moscow, with Kazimir Malevich's white-on-white squares (see p. 58), and other art included Claude Monet's late impressionist painting *Nymphéas* and Pablo Picasso's stage set for Serge Diaghilev's dance production of Manuel de Falla's "The Three-Cornered Hat," Paris.

Films: Abel Gance's pacifist statement, *J'accuse!* (I accuse!), Fritz Lang's *Half Caste*.

Jazz arrived in Europe; other music included Messager's operetta "Monsieur Beaucaire," Birmingham, England.

Lady Astor became the first female member of Parliament, U.K.

Ernest Rutherford split the atom, U.K.

RCA (Radio Corporation of America) was founded, and the first general radio broadcasts began, U.K. and U.S.

Enrico Caruso and Camille Saint-Saëns died.

"Girls Play with Ball" vase by Edward Hald

The glass factory Orrefors was formed in 1898 to take advantage of the sawmill waste that was being dumped by an iron foundry in southeastern Sweden near the island of Öland in the Baltic Sea. At the time Orrefors was making inexpensive utility glass. But in 1916 and 1917, Swedish painters Simon Gate and Edward Hald (1883–1980) were hired to help shift the output to decorative glass. And, indeed, they infused a whole new creative breath of life and dynamism into Orrefors's production, even though they had had no previous experience with glass. The vase shown here reveals the strong influence artist Henri Matisse had had on Hald who studied under Matisse in Paris from 1908–12. It also illustrates the kind of virtuosity he and Gate, working with expert glass blowers and cutters, realized at Orrefors during the 1920s and '30s. Hald's work primarily represented a more graphic, painterly approach than Gate's. This and other special Orrefors vessels were expensive but not one of a kind. Hald and Gate became leaders of Modernism in Sweden and set the standard for other factories that began hiring designers to work with blowers and cutters—in contrast to the traditional practice of relying on blowers and cutters for both design and production. While at Orrefors, Hald also worked in ceramics in the 1920s for other firms.

Date: 1920. Material: Engraved glass. Manufacturer: A.B. Orrefors Glasbruk, Orrefors, Sweden. Photograph: Nationalmuseum, Stockholm, Sweden, inv. no. NM 14/1974.

Events of 1920:

Women were allowed to vote, U.S.

Premiers: The transmission of a regular licensed radio broadcast, by station KDKA, Pittsburgh, Pennsylvania, U.S.; patent of submachine gun ("Tommy gun"), by Thompson, U.S.; League of Nations, Paris; two different models of the electric hair dryer were introduced ("Race" by Racine Universal Motor Company and "Cyclone" by Hamilton Beach Company, both Racine, Wisconsin, U.S.)—the device being patented in 1922.

Architectural Digest magazine was founded, Los Angeles.

Vladimir Tatlin designed the Monument for the Third International, Moscow.

The Duke of Windsor spurred the tennis-shoe craze, during his U.S. visit.

Film: *The Cabinet of Dr. Caligari*, *Pollyana* with Mary Pickford, Buazzoni's *Cesare Borgia*, Marcel Duchamp's first abstract movie.

Literature: Beerbohm's *Seven Men*; Christie's *The Mysterious Affair at Styles*; Colette's *Chéri*; Crofts's *The Cask*, one of the first modern detective stories; Hasek's The *Adventures of the Good Soldier Schwejk* (to 1923); Kafka's *A Country Doctor*.

Music: Ravel's "La Valse"; Oskar Straus's "De Letzte Walzer" (the last waltz), Berlin; Hadley's opera "Cleopatra's Night," New York Metropolitan Opera; first complete performance of Holst's "The Planets," London.

Paul Deschanel became president; Clemenceau resigned; and Etienne Millerand was elected premier, France.

Jean d'Arc was canonized by Pope Benedict XV.

In reparations concerning World War I, U.K. received the Palestine Mandate.

Auguste Renoir, F.W. Woolworth, and Emiliano Zapata died.

French women's clothing designer Gabriel "Coco" Chanel (1883–1971), working with Ernest Baux, was the first couturière to concoct a completely artificial perfume. "Chanel No. 5," a secret formula of more than 80 organic chemical ingredients, was different from previous scents composed of easily identifiable floral elements. However, apocryphal stories have abounded—some fostered by Chanel herself—concerning the perfume's creation. One account suggests that the name "No. 5" concerned its creation on the fifth day of the fifth month of 1921. Another, more believable account proposes that the final formula was the fifth in a series of experiments. The origin of the bottle and the label are also not clear. Possibly both are autonomous designs. The flacon may have been an industrial, generic perfume bottle, and the label, terribly plain for the times, may been been a piece of paper affixed for mere identification, like those used by chemists. Some sources assign the source of the bottle's form to Lou Dofmann of the glass factory Verreries Pochet et du Courval. Dofmann may have designed it as a generic perfume flacon. Chanel suggested that the stopper had been purposefully formed to resemble a cut emerald; more likely, the stopper was an integral part of the bottle as provided by the glass factory.

Design: 1921. Materials: Glass, paper. Manufacturer: Attributed to Verreries Pochet et du Courval, France. Photograph: Permission Chanel, Paris.

Events of 1921:

Marcel Breuer, influenced by the work of the De Stijl group, especially Gerrit Rietveld, made his first chairs, Germany.

Premiers: Radio-parts firm Braun, Frankfurt am Main; "Band-Aid" adhesive bandage, by Earl Dickson; dinnerware in stainless steel (patented in 1916), by the Silver Company, Meridan, Connecticut, U.S.; whistling teakettle, by Joseph Block; domestic iron with an automatic adjustable thermostat, by Joseph W. Myers; cultured pearls; magnetron, an electronic tube to produce microwaves, by Albert W. Hill; polygraphy, or lie detector, by Canadian-American medical student John Augustus Larson; inkblot test, by Hermann Rorschach, Switzerland.

The first of the industrial-design exhibitions, organized by curator Richard Bach, at The Metropolitan Museum of Art, New York.

Einstein Tower, the intricate study in amorphically shaped poured concrete over brick, was built by Eric Mendelsohn, Berlin. A tower was built by vernacular architect S. Rodilla, Watts section, Los Angeles.

Karl Capek, a playwright, coined the word "robot" for the mechanical people in his play *RUR*, Czechoslovakia.

Literature: Dos Passos's *Three Soldiers*, Lawrence's *Women in Love*, Marcel's *La cœur des autres*, O'Neill's play "Anna Christie," Pirandello's play "Sei personaggi in cerca d'autore" (six characters in search of an author), Sabatini's bestseller *Scaramouche*.

The Duke of Westminster sold Gainsborough's *Blue Boy* and Reynold's *Portrait of Mrs. Siddons* to an American for £200,000.

Albert Calmette and Camille Guérub developed the B-C-G (Bacillus calmette-guérin) tuberculosis vaccine. France.

Irish-born Eileen Gray (1878–1976) studied painting and drawing in London, was an assistant in the furniture workshop of D. Charles where she learned about lacquerwork, and settled in Paris in 1907. From about 1910 she began working on decorative panels, screens, and then individual pieces of furniture for rich clients. Trained by Seizo Sugawara in Paris from 1907, she became a master lacquer worker. Her first complete object —the "Le Destin" folding screen of 1919–22 shown here—was made for Parisian couturier Jacques Doucet who had begun collecting distinguished modern French furniture. Gray produced another example of the screen and other furniture, including an extraordinary day-bed called the "Pergola" (canoe) for modiste Suzanne Talbot on the rue Lota in Paris. In the sparsely furnished Talbot apartment, blocks of the lacquered screen were repeated on the walls of the entrance hall. It was Gray's architecturally derived forms that brought her to the attention of the De Stijl group who may have appreciated her more than the French. From her shop, called by the pseudonym Jean Désert, on the rue du Faubourg Saint-Honoré in Paris, she sold one-of-a-kind furnishings from 1922–30, but Gray fell into obscurity after World War II. She was not rediscovered until the 1970s when she was in her 90s and when some of her furniture began to be mass produced.

Date: 1919–22. Materials: Lacquered wood, aluminum. Manufacturer: Jean Désert (Eileen Gray), Paris. Photograph: Virginia Museum of Fine Arts, Richmond, gift of Sydney and Frances Lewis. © Virginia Museum of Fine Arts.

Events of 1922:

Etiquette: The Blue Book of Social Usage by Emily Post was published.

Reader's Digest magazine published its first issue.

Premiers: Skywriting; Waring Blendor, by Stephen J. Poplawski, U.S.; engines for the Zeppelin dirigibles, by Wilhelm Maybach, Germany; "5 CV" auto by Citroën, France; crossing of the South Atlantic by hydroplane, by Cabral and Coutinho; regular radio-program broadcasts by the BBC, London; insulin for the regression of diabetes.

Film: Lang's *Dr. Mabuse*; Perfect Pictures' *The Power of Love*, the first 3-D movie; *Toll of the Sea,* the first Technicolor feature.

James Joyce's novel *Ulysses* was published. British director of public prosecution, Sir Archibald Bodkin, concluded the book was obscene after he read a few dozen pages at the end of a copy that was seized at Croydon airport on 22 December; the book was barred from publication and sale in the U.K. until 1936.

Paul Jaray first presented his aerodynamic auto bodies, Berlin.

Unsuccessfully designed to recover gold from the ocean floor, the German *Meteor* oceanographic expedition was equipped with sonar to measure the ocean depths. (In 1925, the expedition discovered the Mid-Atlantic Ridge.)

Russia became the Union of Soviet Socialist Republics.

Benito Mussolini took power, Italy.

Vitamin E was discovered by Herbert McLean Evans and K.J. Scott; vitamin D, as a treatment for rickets, was discovered in cod liver oil by Elmer McCollum.

Alexander Graham Bell, Marcel Proust, Georges Sorel, and Lillian Russell died.

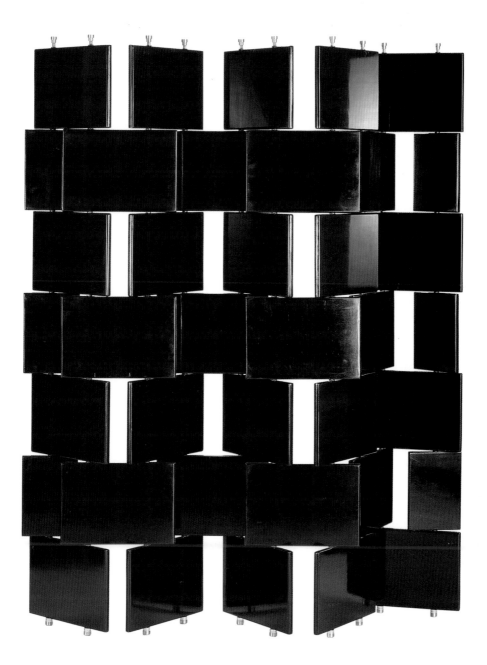

Cup by Kazimir Malevich and Il'ia Chashnik

Kazimir Malevich (1878–1935), best known as an abstract painter, was also active in Russian in textiles, architectural concepts, and ceramics. He had participated in the European art world before the Russian Revolution of 1917. Launched in 1915, his Suprematist compositions, a term he coined, were paralleled to the development of Modernism in Western Europe. It was at the Russian government's State Porcelain Factory, not unlike the French government's Manufacture Nationale du Sevrès, that Malevich realized some interesting three-dimensional ideas. The country was devastated after the Revolution; artistic porcelain at the factory was meagre; and cost for new products was expensive. Therefore, decorations were applied to Czarist-period blanks and included a hodgepodge of styles—from conservative to avant-garde. The playful, somewhat impractical half cup shown here became one of the few completely new dinnerware products. In the Suprematist style, Malevich designed the shape and his former student Il'ia Chashnik (1902–29) rendered the decoration. This peculiar cup and other similar wares, like figurines, were primarily produced for exhibition or sales in the West. Malevich and Chashnik worked at the factory from 1922–24.

Date: 1923. Materials: Porcelain, enameled decoration. Manufacturer: State Porcelain Factory, St. Petersburg, Russia. Photograph: Courtesy Cooper-Hewitt, National Design Museum, Smithsonian Institution/Art Resource/NY, The Henry and Ludmilla Shapiro Collection, partial gift and purchase through The Decorative Arts Association Fund and the Smithsonian Collections Acquisition Program Fund, 1989-41-8.

Events of 1923:

The Charleston, an African-American dance, was adopted by Broadway, New York.

Vers une architecture, espousing theories on the excellence of vernacular and industrial design, by Le Corbusier was published, Paris.

Premiers: The rubber diaphragm as a contraceptive device; Maidenform brassier; Warner Bros. Pictures, Hollywood; Zenith Radio, Chicago; Milky Way and Butterfinger candy bars; *Time* magazine; Hertz Drive-Ur-Self auto rentals; Pan American World Airways; photoelectric cell, or the "electric eye," by Sherbins; mechanical-scanner television, by James Logie Baird, U.K.

Exhibition of the avant-garde architecture, art, and design group De Stijl was organized at Galerie Léonce Rosenberg, Paris.

First Salon des Arts Ménagers was held, Paris.

First Biennale (known as the Triennale di Milano from 1933) was held, Monza, Italy.

Film: *Safety Last* with Harold Lloyd, highest-paid comedian, and *Hunchback of Notre Dame* with Lon Chaney, Hollywood.

BMW Motorcycles was established with the introduction of the "R 23" model, Germany.

The autogiro principle was developed by pilot Juan de la Cierva, Spain.

Dochez, Georges, and Dicks discovered that scarlet fever was caused by streptococci for which they developed an antitoxin.

Sigmund Freud's *The Ego and the Id* was published.

Germany suspended war-reparation payments, and France invaded the Ruhe Valley.

Sarah Bernhardt and Gustave Eiffel died.

Wilhelm Wagenfeld (1900–90) worked in a silver factory and studied at art schools in Germany before entering the silversmithing section of the Bauhaus school in Weimar in 1923 where he eventually became an instructor. While some of his other work—including examples in glass—is exceptional, his most celebrated design is probably the "Bauhaus" lamp, named the "MT 8" at the time of its creation. Wagenfeld's Bauhaus teacher László Moholy-Nagy had asked him to design a lamp in 1923. One of the first examples was given to the school's headmaster, Walter Gropius, as a birthday present in May 1924. Until 1928, the lamp was fitted with either a round metal base with three ball feet or with a thick clear-glass base, and in variant models, one with a tubular-metal stem. Swiss Bauhaus silversmithing student Carl Jacob Junkers (1902–97) claimed that he designed the all-glass version, but there is no documentation to support the assertion. The lamp has become an icon of 20th-century design and an example of the Bauhaus philosophy that fostered rigid geometrical forms appropriate for multiple production. The fixture was produced in numerous, dreadful post-war interpretations until Wagenfeld re-edited the lamp in production today by the Bremen firm Technolumen.

Date: 1923–24. Materials: Chromium-plated metal and mouth-blown glass (early model shown here). Manufacturers: Metalworking workshop, Bauhaus, Weimar and Dessau (from 1924); Schwintzer und Gräff, Berlin (1928–30); Bünte und Remmler, Frankfurt (plagiarized, c. 1930); Dr. Heinrich König, Architekturbedark, Dresden (1931); Imago, Italy (from 1971); others (from 1970s); Technolumen, Bremen (from 1980). Photograph: Courtesy The Museum of Modern Art, New York. Gift of Philip Johnson. Photograph © 1999, The Museum of Modern Art.

Events of 1924:

Architecture included the Schroeder House by Gerrit Rietveld and its owner Truus Schroeder-Schraeder, Utrecht, Netherlands, and La Pravda building by Victor and Alexandr Vesnin, Leningrad.

Ford Motor Company made its 10,000,000th auto, when over half the world's cars were Ford "Model Ts."

Asa G. Candler sold the Coca-Cola Company to Robert W. Woodruff for $25,000,000, Atlanta, Georgia, U.S.

Hilton Hotel Corporation and Bell Laboratories for telephone research were founded, U.S.

Premiers: Kleenex tissues (known at first as Celluwipes); Wheaties dry cereal; Macy's Thanksgiving Day Parade, New York; Plexiglas (trademarked 1934), by Barker and Skinner, all U.S.; patent of the self-winding watch.

"Rhapsody in Blue," a romantically emotional symphony in the jazz idiom, was written by George Gershwin, New York.

Art: Brancusi's The Beginning of the World and Kokoschka's Venice. André Breton published his first surrealist manifesto, Paris.

Films: Clair's Entr'acte with Francis Picabia and Murnau's Nosferatu.

British Olympic runners (featured in Hudson's Chariots of Fire film of 1981) were outfitted by Reebok shoes.

Dassler Schufabrik (sports shoes) opened, Germany. (Its shoes were worn by Jesse Owens in 1936 Berlin Olympics.)

Iconoscope, an early type of television, was patented. by Russian-American physicist Vladimir Kosma Zworykin, U.S.

Photographs were transmitted by radio, from New York to London.

Ferruccio Busoni, Joseph Conrad, Franz Kafka, V.I. Lenin, Giacomo Puccini, and Louis Henri Sullivan died.

From the early 1920s, Hungarian Marcel Breuer (1902–81) studied and then taught at the German Bauhaus architecture, design, and art school, where he began working on tubular-steel furniture. Before developing his now virtually ubiquitous "B32" cantilevered chair in wood, cane, and tubular steel, he created an abstraction of the highly padded armchair used in exclusive English clubs for men; it was initially called the "Club." It soon became known as the "Wassily" chair, so named in honor of, but probably not designed specifically for, his friend, Russian artist Vassily Kandinsky, a resident at the Bauhaus at the time. A plumber helped Breuer to weld together the first model, made from metal tubes acquired from Mannesmann, the huge German steel works that invented seamless extruded steel tubing. At first, the straps of the "Wassily" were of a fine horsehair material which for serial production was replaced by *Eisengarn*, or Ironcloth, a tough twill fabric used for military belts and the like. The chair is made by a number of firms today, in varying levels of quality, with either cotton canvas or leather straps. Revised a number of times, the current production model, including the example shown here, is a version of late 1927 or early 1928.

Date: 1925. Materials: Chromium-plating over nickel-plated tubular steel and canvas (early example shown here). Manufacturers: Standardmöbel Lengyel, Berlin, Germany; DESTA (Deutsche Stahlmöbel), Berlin (from 1929); Gebrüder Thonet, Frankenberg, Germany (from early 1930s); Gavina, Foligno, Italy (from c. 1960–68), Knoll Associates, East Greenville, Pennsylvania, U.S. (from 1968); others. Photograph: Cooper-Hewitt, National Museum of Design, Smithsonian/Art Resource, NY. Photo: Gary Laredo, 1956-10-1.

Events of 1925:

Exposition internationale des arts décoratifs et industrielle moderne opened to great fanfare, Paris.

William L. Murphy invented the Murphy bed, a type that folds into an upright position against a wall, U.S.

Bella Barenyi designed what became the Volkswagen, was too poor to patent it, but later sued Porsche and won.

Chrysler Corporation was founded by Walter P. Chrysler.

Premiers: The cloche hat and the straight dress above the knees and without a waistline; contract bridge game, by Harold S. Vanderbilt; the motel (the Milestone Motel), Monterey, California; *The New Yorker* magazine; analog computer (to solve differential equations), by Vannevar Bush and co-workers; commercial availability of chromium; industrially developed synthetical oil, Germany; prediction of sound movies, by A.O. Rankine; Chicago-style jazz in Europe; Cinemascope (wide-screen movies).

Film: Eisenstein's *Battleship Potemkin*, based on the 1905 mutiny, Russia; *The Gold Rush* with Charlie Chaplin; Pabst's *The Joyless Street*; Clair's *The Ghost of the Moulin Rouge*.

U.S. singer-dancer Josephine Baker arrived in Paris.

Literature: Fitzgerald's *The Great Gatsby*, Coward's play "Hay Fever," Kafka's posthumously published novel *The Trial*.

Adolf Hitler recognized the Nazi Party's 27,000 members and published the first volume of *Mein Kampf*, Germany.

First television transmission of recognizable human features, by John Logie Baird, U.K.; Vladimir Kosma Zworykin applied for color-television patent, U.S.

First solar eclipse in 300 years occurred in New York.

John Singer Sargent and Erik Satie died.

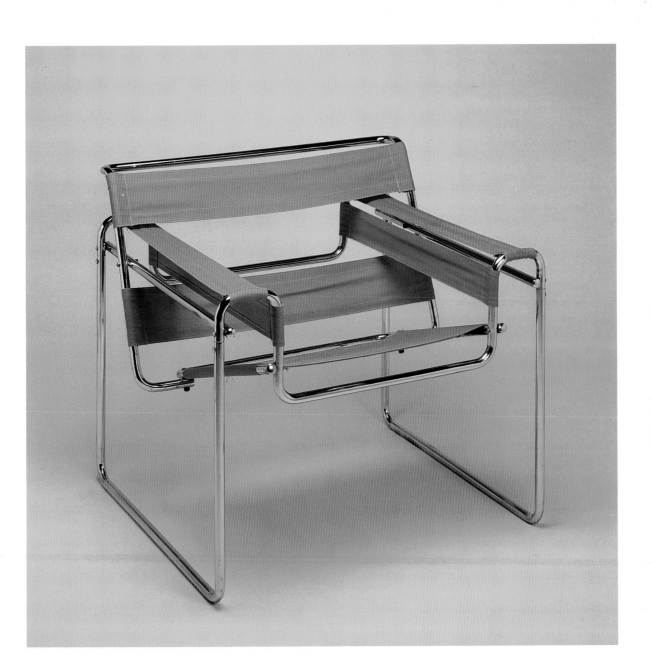

"B33" side chair by Mart Stam

Harry E. Nolan's U.S. patent of 1922 for a canti-levered chair and Gerhard Stüttgen's configuration of 1923 have long since been forgotten. The examples we remember today are a version by Dutch architect Mart Stam (1899–1986), the ever-present "B32" of 1927–28 by Marcel Breuer, and the "MR" of 1927 by Ludwig Mies van der Rohe. They all stemmed from the availability of flexible steel tubing developed by Mannesmann (see p. 62). Stam's first, awkward version of 1924, whose solution relied more on architecton-ics than technology, was made of rigid cast iron assembled with plumbing joints. Mies, an instruc-tor at the Bauhaus school, informed Stam of the bending process that Breuer was using at the school. In 1926, Stam adopted the technique, refined his cantilevered chair, and, on Mies's invitation, showed it at the Weissenhof Housing Exhibition in Stuttgart in 1927. It was immediate-ly put into production as the "B33." For this final version, a tubular member between the front legs was replaced by a spacing rod under the seat, and the back was tilted. And, weeks after Stam's chair appeared at the exhibition, Mies introduced his own version. Lacking clear attribution, con-tentious lawsuits among Hungarian businessman Anton Lorenz, Thonet, and other manufacturers from the late 1920s concerned sole rights to the cantilever principle.

Date: 1926. Materials: Chromium plating over nickel-plated steel tubing, fabric (shown) or leather. Manufacturers: L.&C. Arnold Eisenmöbelfabrik, Schorndorf, Germany (from 1927); Gebrüder Thonet, Frankenberg, Germany (from the early 1930s; today as the "S33") (example shown here of the 1930s). Photograph: Courtesy Gebrüder Thonet.

Events of 1926:

Alexander Calder made his first wire sculptures.

Container Corporation of America was founded and became the first to develop the first cardboard packaging.

Auto production was at 4,000,000, with 43 companies mak-ing them, U.S.

Premiers: The pop-up toaster, by Charles Strite and Murray Ireland, U.S.; production of a domestic iron with an automatic adjustable thermostat, invented by Joseph W. Myers in 1921, U.S.; steam iron, first sold in New York for $10; Café Unie, by J.J.P. Oud, Rotterdam; built-in kitchen, by Margarete Schütte-Lihotzky, Frankfurt; industrial-design firm (first client Eastman Kodak) of Walter Dorwin Teague, New York; television images of moving objects, by John Logie Baird, U.K.; flight over the North Pole, by Richard Byrd; discovery of dinosaur eggs, by Roy Andrews, Gobi Desert; Prestone antifreeze for auto radia-tors, U.S.

House of Tristan Tzara, an initiator of the Dada art move-ment, was designed by Adolf Loos, Paris.

First liquid-fuel-propelled rocket (9-lb. thrust) was launched by Robert Goddard, U.S.

Film: Lang's *Metropolis* concerned the frightening dehu-manization potentially resulting from industrialization, Germany. *The Jazz Singer* with Al Jolson introduced talking motion pictures, Hollywood.

Sears, Roebuck & Company, known for its mail-order catalogue ordering system, opened its first retail store, U.S.

Albert Michelson measured the velocity of light within 0.00001% error.

Russian director Stanislavsky's book *An Actor Prepares* went on to transform 20th-century acting techniques.

Antonio Gaudí, Harry Houdini, and Rudolph Valentino died.

Little was known of Edouard-Wilfrid Buquet or his adjustable lamp until some research in the 1970s unearthed the attribution and the patent date of 1927. The fixture had become one of the most practical and popular lighting fixtures soon after its appearance in 1929 on the "Desk of a Technician" by Lucien Rollin at a Paris exhibition. The lamp was widely published in French magazines of the late 1930s. It was used by a number of designers, like Marcel Breuer, but particularly by French ones, such as Joubert and Petit, Maurice Barret, Louis Sognot, and Marcel Coard. There were subsequent versions of a human-arm-like lighting fixture: the "Anglepoise" of 1932 designed by British automobile engineer George Carwadine and the "Luxo" of 1937 by Norwegian designer Jacob Jacobsen. Unlike these models that used springs to hold the angled arm stable, Buquet's balancing principle was made possible by a heavy weight at the end of the arm that would "occupy all positions and orientations in space without exception, because the arms can inscribe complete circles around each of their joints," according to the description in Buquet's granted patent. The adjustment knob on the top of the reflector was necessary because the reflector offered little ventilation and became very hot to the touch.

Date: Patented 1927. Materials: Nickel-plated brass, aluminum, lacquered wood. Manufacturer: W.W. Buquet, France. Photograph: Courtesy Die Neue Sammlung, Munich, Germany, inv. no. 115/75. Photo: Angela Bröhan, Munich.

Events of 1927:

Ludwig Mies van der Rohe designed his version of the cantilevered chair, based on the principles of Mart Stam's "B 33" that was included in the Weissenhof Housing Exhibition of the same year, Stuttgart, Germany.

Decorative paper was introduced by Herbert Faber and Daniel O'Connor into their plastic laminate Formica, U.S.

Premiers: Volvo autos, Sweden; all-electric jukebox; homogenized milk, by Borden; Gerber baby food; automatic traffic light, London; "Universal" typeface, by Herbert Bayer, Germany; Cassina furniture factory, Italy; Cranbrook Academy of Art, Bloomfield Hills, Michigan, U.S.; mandatory auto license plate, Massachusetts, U.S.; underwater color photography (in *National Geographic* magazine); trophy of the Academy Award of Motion Picture Arts and Sciences, by MGM art director Cedric Gibbons, Hollywood; transatlantic telephone link.

Abel Gance's film *Napoléon* was shown at the Paris Opéra.

Mickey Mouse, named by Walt Disney's wife, made his début in the animated black-and-white film *Steamboat Willy*.

American pilot Charles A. Lindbergh flew non-stop and solo in the Ryan monoplane, *Spirit of St. Louis*, from New York to Paris in 33 1/2 hours.

Verein für Raumschiffart (Society for Space Travel), inspired by ideas concerning space and rockets from Hermann Oberth's 1924 book, was founded, Germany; members included Wernher von Braun (a leader of Germany's rocket program, subsequently head of U.S. rocket program) and Willy Ley (later himself author of books that made rocket technology understandable to U.S. lay readers).

Isadora Duncan and Gaston Laroux died.

Chaise longue by Le Corbusier, Jeanneret, and Perriand

Swiss-French architect Le Corbusier (1887–1965) with his cousin Pierre Jeanneret (1896–1967) and Charlotte Perriand (b. 1903) designed the adjustable chaise shown here. It first appeared in the Villa Church of 1929 in Ville d'Avray, France. Afterwards, in the same year, the public first saw it at the Salon d'Automne in Paris. As Le Corbusier described his houses as "machines for living," thus he called this chaise a "real machine for rest." And he considered industrial design to be on a par with painting. However, his dicta were jarring to many in the 1920s. Reinterpreting the bentwood chaises of Thonet, the French designers transformed bentwood into bent tubular steel and made the base into a separate unit for their version of the chaise here. Another progenitor, "Le Surepos" by Dr. Pascaud, may have served as further inspiration. An exotic example with a straight back edge and a pad covered in leopard skin was created for the interior scheme of 1929–33 arranged by Eckart Muthesius (the son of Hermann Muthesius) and Klemens Weigel for the Maharajah of Indore's palace in India. The standard model of the chaise is produced today by a number of manufacturers.

Date: 1928. Materials: Chromium plating over nickel-plated steel, fabric, leather, rubber. Manufacturers: Thonet France, Paris (from 1928); EMBRU-Werke (Eisen- und Metallbetten-fabrik), Rüti, Switzerland (1930s); Heidi Weber, Switzerland (from 1959); Cassina, Italy (from 1965); others. Photograph: Courtesy The Museum of Modern Art, New York.

Events of 1928:

Domus design and architecture journal, conceived and edited by Gio Ponti, published its first issue, Milan, Italy.

Hannes Meyer was named director of the Bauhaus architecture, art, and design school, Dessau, Germany.

Opel "Rak II" became the first auto to be propelled by a rocket engine, reaching 280 km/h. on the Avus-Berlin route.

Premiers: Commercial bubble gum (Fleer's Double Bubble); Scotch-brand cellophane adhesive tape, by 3M, U.S.; hydro-genated, homogenized peanut butter, by Peter Pan; short-wave radio transmission between England and U.S.; radio beacon; polarizing filter, by Edwin Land; quartz-crystal clock, by J.W. Horton and W.A. Harrison; Pap test for diagnosing uterine cancer, by Greek-American George Papanicolau; Columbia Broadcasting Company, by cigar-company executive William S. Paley, U.S.; broccoli introduced into U.S. from Italy; founding of the Congrès Internationaux de l'Architecture Moderne (CIAM), Switzerland; transatlantic television transmission, England to America, by John Logie Baird.

Based on British playwright John Gay's "Beggar's Opera" of 1728, Brecht's musical "The Three Penny Opera" was the biggest success of the theater-crazed Weimar era, Berlin.

Margaret Mead's *Coming of Age in Samoa* described the passage from childhood to adulthood in a primitive society.

Scottish bacteriologist Sir Alexander Fleming discovered penicillin through a chance exposure to staphylococci, but penicillin was not used as a therapy until 1942 in the U.S.

Bergdorf Goodman women's clothing store was opened, 58th Street and 5th Avenue, New York.

Roald Amundsen, Leos Janacek, and Thomas Hardy died.

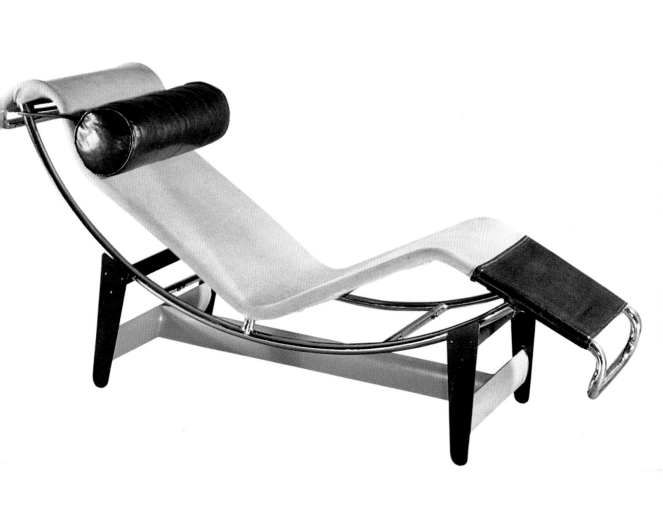

"Barcelona" lounge chair by Ludwig Mies van der Rohe

Ludwig Mies van der Rohe (1886–1969) was the son of a stonecutter in Aachen, Germany, and eventually worked in the Berlin office of Peter Behrens (see p. 30) with others who became pioneers of International Style architecture. In 1913 he added his mother's maiden name "Rohe" to his own to make it more distinguished. Highly active in architectural circles, he directed the Weissenhof Housing Exhibition in Stuttgart where Mart Stam's cantilevered chair (see p. 64), as well as his own version, was introduced. Mies had designed furniture for his early houses. So, likewise, at the 1929–30 world exposition in Barcelona, Spain, Mies included "Barcelona" chairs, as they have come to be known. When the King and Queen of Spain visited the pavilion to sign the guest book they were to sit on the chairs as thrones, but they did not. Examples of the chair also appeared in Mies's Tudendhat house of 1929 in Bruno, Moravia. The first copies were made by the locksmith Joseph Müller. The leather straps beneath the cushion are for appearance only; stiff rubber strips beneath the rubber offer the real support. The design of the cushions was probably by Mies's associate Lily Reich, a frequent collaborator and a teacher at the Bauhaus where Mies was its last director from 1930 before the Nazis shut it down in 1933.

Design: 1929. Materials: Chromium-plated steel, leather, rubber. Manufacturers: Berliner Metallgewerbe Joseph Müller, Neubabelsberg, Germany (from 1929–31); Bamberg Metallwerkstätten, Berlin (from 1931); Knoll Associates, East Greenville, Pennsylvania (from the 1940s); since, by others. Photograph: Courtesy The Museum of Modern Art, New York.

Events of 1929:

Stock Market crashed on 29 October, New York, and marked the beginning of the Great Depression worldwide.

Braun electronics began producing radios and phonographs and became one of the first to incorporate speaker and receiver into one set, Frankfurt, Germany.

The Museum of Modern Art was founded by Alfred H. Barr Jr., Blanchette Rockefeller, and others, New York.

Villa Savoie was built by Le Corbusier, Poissy, France.

Premiers: "Tin Tin" comic strip, France; home-applied hair coloring; mobile home (trailer); Susie Cooper Pottery, in a factory rented from Doulton, England; foam rubber, by Dunlop Rubber Company, U.K.; Kölner Möbelmesse (Cologne furniture fair), Germany; FM radio; electroencephalograph (EEG), by psychiatrist Hans Berger, Germany; insulin shock for the treatment of schizophrenia, by Manfred J. Sakel; front-wheel drive in autos; "Popeye the Sailor Man" comic strip, U.S.; nudist colony in New Jersey, U.S.; round-the-world flight, by the *Graf Zeppelin*, 19,000 miles in 21 days, 7 hours; flight over Antarctica, by aviator Richard E. Byrd; 16-mm color film, by Kodak.

"Hubble's constant" or "Hubble's law," announced by Edwin P. Hubble, explained expansion of the Universe, U.S.

Lateran Pacts made Catholicism Italy's state religion; Vatican City became an independent country.

First large-scale attack on Jews by Arabs, Jerusalem.

Gangster Al Capone's gang killed six members of a rival gang and one bystander; because of the date, the incident was named the St. Valentine's Day Massacre, Chicago.

Georges Clemenceau, Sergei Diaghilev, and David Buick died.

In the 1920s and '30s in Italy, the opposites of traditionalism and Modernism tugged at each other. Nevertheless, a desire to modernize was being enthusiastically sought by industrialists. American mass-manufacturing techniques and product standardization were being emulated. The transition was officially signalled in 1930 when the international biennial in Monza began including "decorative and modern industrial art," rather than only "decorative art." The year also marked the introduction of Alfonso Bialetti's "Moka Express" coffee pot. Made according to new industrial techniques, the first pot was, nevertheless, an expensive, handmade object in a squat, somewhat unrefined form—far different from the sleek version mass-produced from 1933 which has remained unchanged. Today the "Moka Express" can be found in 95% of Italian households and 250,000,000 homes worldwide. But it has only been popular since the 1950s when it was marketed by Alfonso Bialetti's brother Renato. Alfonso had been inept at selling the product. The eight flat planes offer great resistance, without bursting, to the high pressure of the steam required for a good cup of espresso. In the bottom section, the steam forces water through the coffee grounds and spews into the top where the dark liquid is captured and retained. The pot, heated on a stove's open flame, is acceptable for use at the dinner table, though the taste of the graphics is questionable.

Design: 1930–33. Materials: Chill-cast aluminum (bottom), die-cast aluminum (top), heatproof Bakelite (handle and knob). Manufacturer: Bialetti, Omegna, Italy (current model shown here). Photograph: Courtesy Bialetti.

Events of 1930:

The 50,000,000th unit of the "No. 14" chair of *c.* 1855 was produced by Thonet, Germany.

Stainless steel for kitchen surfaces was first promoted at the Stockholm Exhibition, Sweden.

Deutsche Werkbund exhibition was part of the Salon of the Société des Artistes Décorateurs whose French members, reacting against the conservatism of the S.A.D., founded the Union des Artistes Modernes (U.A.M.) in 1929 and first showed their work the following year, Paris.

American Union of Decorative Artists and Craftsmen (AUDAC) was founded by Paul Frankl and others, New York.

Premiers: Pininfarina auto body-styling studio, Turin; Berta Stadium, by Pier Luigi Nervi, Florence, Italy; extension of electricity to the countryside, France; sliced bread, U.S.; tape recorder using magnetized plastic tape, Germany; jet engine, patented by Frank Whittle, who also perfected the turboreactor this year, U.K.; polystyrene, by I.G. Farberindustrie, Germany; PVC (polyvinylchloride), by W.L. Semon at B.F. Goodrich Company, U.S.; freon gas for refrigerators, by Thomas Midgley Jr. (first produced 1931 and much later associated with destruction of the ozone layer); automobile radio, by Paul Galvin of Galvin Corp.(later named Motorola).

Vannevar Bush completed the differential analyzer, or an analog computer, U.S.

The planet Pluto was discovered by C.W. Tombaugh, U.S.

William Beebe and Otis Barton dove to 417 m (1,368 ft.) in a bathosphere, a new type of submersible.

Wernher von Braun founded an organization for space travel with a rocket-launch site near Berlin.

Constantinople, formerly Byzantium, was renamed Istanbul, Turkey.

Arthur Conan Doyle, D.H. Lawrence, and Vladimir Mayakovsky died.

René Herbst (1891–1982) was the ring leader of those fed up with the conservatism of the members of Société des Artistes Décoratifs who had shown their work at the annual events in Paris since 1904. He and others left S.A.D. to form the Union des Artistes Modern (U.A.M.) in 1929. A curmudgeon, he became one of the radical initiators of the use of industrial materials in the decorative arts in France. Today Herbst is best known for a series of chairs from 1928 that incorporated what the French call *sandows*, or elastic bungee cords; the word literally means "shock absorber" or "expander." They were first shown in 1929 at the annual Salon d'Automne venue. Unlike overstuffed chairs, the chairs were informal and compatible with the sparsely furnished small rooms that people were beginning to occupy, and their transparency lessened their physical presence. A prolific designer, he was commissioned by the Prince Aga Khan to lay out and furnish a suite in Paris; it was to illustrate that modern materials and design could be used successfully to create an elegant space. Herbst also produced a number of one-of-a-kind pieces for himself and his own apartment in Paris such as the heavy forged-steel bookcase/divan unit shown here. The colorful geometric motif was handpainted by Herbst himself on the doors of the floating cabinet; a mattress was originally placed on the frame behind the canbinet; and the clear-glass and steel shelves are adjustable. Herbst's interest in and use of metal together with an irascible personality earned him the title "the iron man."

Design: 1930–31. Materials: Forged steel, lacquer, glass, wood. Manufacturer: Establissement René Herbst, France. Photograph: Courtesy Galerie de Beyrie, New York.

Events of 1931:

Exposition coloniale opened, Paris.

Tubular-steel furniture company PEL, U.K., and Den Permanente store, Copenhagen, were founded.

First television in France was broadcast by René Barthélemy.

Inspector Jules Maigret, the persistent pipe-smoking detective, was created by Belgian-born author Georges Simenon, Paris.

Premiers: Rationalism in design and architecture, Spain; Clairol hair dye, France; electric razor, by Schick; telex, by Bell Laboratories; board game Criss Cross (name changed to Scrabble in 1948), by Alfred Butts, U.S.; electron microscope, by Max Knoll and Ernst Ruska, Germany; baffle ball (later called pin ball), by David Gottlieb, U.S.; Alka-Seltzer stomach settler, U.S.; "The Star Spangled Banner" as the U.S. national anthem; flight around the world in 15 hours, 51 minutes, by Wiley Post and Harold Gatty; large balloon into the stratosphere, by Swiss physicist Piccard and associate Kipfer.

Pearl Buck wrote *The Good Earth* in a style derived from the King James version of the bible and Chinese sagas.

George Washington Bridge was opened, with double the span (1,006 m) of any to date, New York.

Winkler launched the first European liquid-propelled rocket.

Growth of viruses in eggs by Ernest Goodpasture made possible the production of vaccines, one of which became effective against polio.

Ernest Orlando Lawrence developed the cyclotron, the first workable particle smasher.

Christ statue atop Corcovado mountain was dedicated, Rio de Janeiro, Brazil.

Thomas Alva Edison, Thomas Lipton, and Anna Pavlova died.

Called "an American classic" and "one of the best products ever made," the famous cigarette lighter shown here with its distinctive opening and closing clicking sound was created by George G. Blaisdell (1895–1978) of Bradford, Pennsylvania, where his factory is still located today. Lore purports that Blaisdell got the idea for a windproof lighter he had seen at a party. He was chatting with someone who was using an awkward, unattractive Austrian lighter. Blaisdell suggested that, since the man was well dressed, he should have "a lighter that looks decent." The response—"but it works!"—impressed Blaisdell, and, in 1932, he introduced his version, named the "Zippo," a play on "zipper." The patent date of 1936 has caused some confusion about its 1932 début. The hinge on Blaisdell's first version shown here was soldered to the outside of the case, but, from 1936, it was fused to the inside wall with only the cylinder revealed when closed. This first model was about $1/4$ in. (5 mm) taller than today's version which has rounded corners. Over the years, "Zippos" have been engraved with insignia and also been made in slim sizes, but the mechanism has remained essentially unchanged. About 300,000,000 "Zippos" had been sold by 1999 and, if laid end to end, would reach half-way around the world. If a lighter breaks, the factory will repair it free—forever.

Date: 1932. Materials of the original example of 1933 shown here: Seamless brass tubing by Chase Brass & Copper, Waterbury, Connecticut; polished or matte-finished chromium plating. Manufacturer: Zippo Manufacturing Company, Bradford, PA, U.S. (from 1933). Photograph: Zippo Manufacturing Company.

Events of 1932:

International Style exhibition, the event that first recognized and named the movement, was held at The Museum of Modern Art, New York.

Alexander Calder first exhibited his "mobiles" (coined by Marcel Duchamp) and "stabiles" (coined by Jean Arp). (Mobile in French means both "movement" and "motive.")

Broadcasting House was built by architects Meyer and Hand and featured Eric Gill's sculpture *Prospero and Ariel*, London.

Premiers: The design of streamline trains, designed by Henry Dreyfuss, Raymond Loewy, and Norman Bel Geddes, independently, U.S.; opening of the Empire State Building (tallest at the time at 102 storeys), New York City; MARS architecture group, London; Metropolitan Cathedral, by Edwin Lutyens, Liverpool; production of the balloon tire for farm tractors, U.S.; Skippy peanut butter, U.S.; London Philharmonic, founded by Beecham; woman flying solo (Amelia Earhart); Revlon cosmetics firm; Radio City Music Hall, New York; gasoline tax, U.S.

"April in Paris" and "Night and Day" were popular songs.

Film: *M*, Germany; *Grand Hotel* won the Academy Award, U.S.; Johnny Weissmuller appeared in his first Tarzan film, and Shirley Temple also made her first film, *Red-Haired Alibi*.

Famine occurred, U.S.S.R.

Literature: Céline's (Louis Detouches's) *Voyage au bout de la nuit* (condemned as vulgar), Maurois's *Le cercle de famille*, Pasternak's poem "Second Birth," Undset's novel *Ida Elisabeth*.

Charles and Anne Lindberg's baby was kidnapped, U.S.

Lady Gregory, André Maginot, and Florenz Ziegfeld died.

The "VE 301 Volksempfänger" (people's radio) was a functional design in a modern material, Bakelite. The severely geometrical cabinet by Walter Maria Kerstling (1889–1970) reveals Art Déco influences. Even though Kerstling's radio was first produced as an independent prototype in 1928, it was only slightly changed when it was subsidized by the National Socialists for mass production from 1933. 12,500,000 radios of this model, or similar, by Kerstling had been sold by 1939; it was an unprecedented number for Europe. The name of the apparatus is revealing of its true purpose. "Volks," or "people," in this case, does not mean everyone; it means "German people." And German people could only receive local radio stations, thus eliminating news from elsewhere that may have been offensive to the Nazi government. It served as a machine to broadcast information issued by radio companies under the control of a Nazi organization, the Ministry for Public Enlightenment and Propaganda. The model number "VE 301" refers to the day, 30 January, when Hitler became the German chancellor in 1933. Further, there is a subtle swastika pattern woven into the grille cloth. Ironically, the design speaks more for Modernism and the aesthetic thrust of the Bauhaus, which the Nazis closed in the year when this radio was put into production, than for the retrograde-nationalistic, pseudo-classical aesthetics which Hitler sought to promote in the arts.

Date: 1933. Materials: Bakelite, other plastics, fabric, metal. Manufacturer: Hagenuh, Kiel, Germany. Photograph: Courtesy Die Neue Sammlung, Munich, Germany, inv. no. 113/98. Photo: Angela Bröhan, Munich.

Events of 1933:

A Century of Progress Exposition opened in Chicago, with turnstiles by John Vassos, theatrically designed restaurants by Norman Bel Geddes, a rotunda for Ford, and The House of Tomorrow in glass and steel with an airplane garage by George Fred Keck (38,000,000 visitors).

Bauhaus architecture, design, and art school was closed by the Nazis, Berlin.

Black Mountain College was founded by John Andrew Rice and Theodore Dreier as a progressive co-educational liberal-arts institution, Asheville, North Carolina, U.S.

Premiers: Air France; product-design department of Montgomery Ward, U.S.; flight of the Boeing "247" airplane; prototypes of the Volkswagen, later known as the "Beetle," by Ferdinand Porsche, Germany; London Underground map, by Henry Beck; Monopoly game, by Charles B. Darrow, U.S.; drive-in movie, Camden, NJ, U.S.

Literature: Benedict's *Patterns of Culture*, to become a classic of comparative anthropology; Eddington's *The Expanding Universe*, one of the first books to popularize modern cosmology; Malraux's *La condition humaine*, a dramatic meditation on destiny. *News-Week* (later *Newsweek*) magazine débuted, U.S.

The Reichstag in Berlin was burned by the Nazis who seized power in Germany. Adolf Hitler, named chancellor of the country and an artist by training, adopted the new swastika trademark for the symbol of the Nazi Party and designed the flag.

Prohibition against alcohol drinking ended, U.S.

William Beebe and Otis Barton dove a record ocean depth of 3,038 ft. (1,001 m) in their tethered bathysphere.

Albert Calmette, King Faisal I of Iraq, F. Henry Royce of Rolls-Royce, and Louis Comfort Tiffany died.

French-born Raymond Loewy (1893–1986) studied electrical engineering in Paris before beginning a 60-year-long career that started when he settled in New York in 1919. From the 1930s, Loewy was assisted by a large staff who produced a large body of work, including the design of automobiles, refrigerators, locomotives, packaging, and trademarks for a number of big, prestigious firms. Furthered by shameless braggadocio, he became the best-known industrial designer in the world. Portraits of him were published on the covers of *Time* and *Der Spiegel* magazines. Loewy's autobiographical *Never Leave Well Enough Alone* of 1951, published in a number of languages, offered reminiscences and dogmatic insights while making no pretense of being a systematic study. Respect for Loewy's "styling" has ebbed and flowed over the years. The irony of his fame lies in the popularity of this never-manufactured, never-copied pencil sharpener, far better known in professional design circles than other examples of his mass-made objects. The device, patented in 1934, superficially followed the principles of streamlining which were based on scientific aerodynamic studies that had begun after World War I and, in the 1930s, symbolized an optimism about the future. Even though the water-drop shape minimized wind resistance, the example here only serves form, not function.

Date: Prototype 1934. Materials: Chromium-plated metal, possibly ivory or plastic. Manufacturer: Unknown.
Photograph: © Rights served.

Events of 1934:

Walter Gropius, former head of the Bauhaus architecture, design, and art school, moved to London, escaping the Nazis.

Effective fluorescent light tube (produced from 1936) was created by Dr. Arthur Compton, a General Electric scientist, resulting in cheaper and cooler lighting than Edison's half-century-old incandescent bulb.

Premiers: The laundromat, Fort Worth, Texas; pipeless organ, by Hammond; secret Swiss bank accounts (Bank Secrecy Law); streamline auto ("Airflow"), by Chrysler, U.S.; development of liquid-fueled rocket, by Wernher von Braun, Germany; Nylon, by Wallace Hume Carothers, U.S.; splitting of the nuclei of uranium atoms, by Enrico Fermi and others; *Normandie* and *Queen Mary*, largest ocean liners to date.

General strike was staged throughout France.

Gandhi suspended the civil disobedience campaign, India.

Hitler was voted by a plebiscite as Der Führer, Germany.

Film: Lubitsch's *Design for Living* and Korda's *The Private Life of Henry VIII.*

Literature: Anouilh's play "La sauvage," Hilton's *Good-Bye, Mr. Chips*, Graves's *I, Claudius* and *Claudius the God*, Dewey's *Art as Experience*, Australian-born Shakespearean actress P.L. Travers's *Mary Poppins.*

Music: Hindemith's symphony "Mathis der maler," Rachmaninov's "Rhapsody on a Theme of Paganini," Stravinsky's ballet-mime "Persephone."

Osovakhim piloted a balloon to 13 miles high, U.S.S.R.

F.B.I. killed "Public Enemy No. 1," John Dillinger, and Texas Rangers killed thieves and murderers Bonnie and Clyde, U.S.

Dionne quintuplets were born, Canada.

Grossglockner Alpine Road was opened to traffic, Austria.

Marie Curie, Edward Elgar, and Fritz Haber died.

The "Hermes-Baby" typewriter was developed from 1932, when it was patented, and sold from 1935. If it were the first truly portable typewriter, it was so by a few moments because the "Studio 42" portable machine by Figini, Pollini, and Schawinsky was introduced by Olivetti in the same year. The Swiss version was more integrally conceived than Olivetti's. The case of the "Hermes-Baby" strengthened the body; the body enveloped the mechanics, particularly the ribbon spools; and the condensed mechanics offered real portability. Giuseppe Prezioso, an engineer at the now-defunct Paillard firm in Switzerland, was an adept mechanic with an eye for rational design. His concept for a low-profile machine incorporated a new system for printing capital letters that saved space inside the body. When a typist pressed an outside key, the carriage was raised, and a capital letter printed when the key for any letter was struck. The aesthetics of other Paillard products—like radios, phonographs, cameras, and photo equipment contemporary to the "Hermes-Baby"—were not as sophisticated. The wings on the helmet and ankles of the Greek god Hermes in the trademark insinuated travel (or portability) and connoted commerce (or the new business class), and, of course, the name "Baby" suggested smallness. Prezioso's typewriter was conceived in the aftermath of Swiss-French architect Le Corbusier's assertion in the late 1920s that machinery was certainly art.

Date: 1935. Materials: Lacquered steel body. Manufacturer: Ernest Paillard & Cie, Yverdon, Switzerland. Photograph: Courtesy Museum für Gestaltung Zürich, Zürich, Switzerland. Photo: Franz Xaver Jaggy.

Events of 1935:

Exposition universelle et internationale de Bruxelles opened, Brussels (20,000,000 visitors).

Russel Wright's furniture in "blonde" wood (a name conjured by his wife for bleached maple) was introduced, to become very popular, U.S.

Brassiere cup sizes, A to D, were made available.

Premiers: Citroën "2 CV" (Deux Chevaux) auto, by André Lefèbvre (prototypes in 1939, production from 1948); Phillips screw, by Henry F. Phillips, U.S.; domestic food disposal, by General Electric, U.S.; beer can (Kreuger Beer), New Jersey, U.S.; portable dishwasher, by Westinghouse, U.S.; permanent design collection of The Museum of Modern Art, New York; design departments of Sears, Roebuck & Co. and of Hotpoint, U.S.; Industrial Design Partnership, by Misha Black and others, London; radar, by R.A. Watson-Watt, U.K.; Toyoda (later Toyota) auto, by Sakichi Toyoda, Japan; rocket to exceed the speed of sound, by Robert Goddard, U.S.; scale for measuring earthquake magnitudes, by C.F. Richter; parking meter, Oklahoma City, U.S.

Olivetti's "Studio 42" typewriter established the typeform of modern typewriters worldwide for the next 40 years, Italy.

"H-1" airplane piloted by its designer Howard Hughes set an air speed record of 352 m./h., U.S.

Musical "Porgy and Bess" by George Gershwin opened to mostly poor reviews, some anti-semitic, New York.

Antonio Caetano de Abreu Freire Egas Moniz developed the prefrontal lobotomy as treatment for mental illness, Portugal.

Konrad Lorenz published a general study of the social behavior of animals, Austria.

Persia became known as Iran.

Alvar Aalto (1898–1976) studied architecture in Helsinki where he set up a practice in 1923 after traveling widely in Central Europe, Scandinavia, and Italy. He became quite active as an architect, town planner, and designer and produced a large body of furniture from 1927, most still in production. But it is his glassware and particularly the "Aalto" vase that stand out above all else among Aalto's work. The vase design won a 1936 competition launched by the Karhula works in Finland for glass to be included in its display at the Finnish Pavilion of the Paris exposition of 1937. Aalto named his competition entry "Eskimoerindens skinnbuxa" (Eskimo woman's leather pants). After various failed production trials, glass was blown into carved-wood molds. In 1954, when the factory began using cast-iron molds, the rippled surface became smooth. The vase was named "Savoy" after Aalto's Helsinki restaurant where it was used, but the staff of Artek, Aalto's furniture firm, soon changed the name to "Aalto." The shape of the vase broke the traditional standard of symmetrically shaped glass vessels. Undulating lines had already appeared in Aalto's architecture, and the side section of his "Paimio" bentwood chair of 1931–32 foretold the vase's curves. Many have speculated on the sources for Aalto's waves. Since *aalto* in Finnish means "wave," the obsession may have been from birth.

Date: 1936. Materials: Blown glass. Manufacturers: Iittala Lasi, Iittala, Finland; Hackman Designor Oy Ab, Iittala, Finland (currently). Photograph: Courtesy Hackman Designor Oy Ab.

Events of 1936:

Exposition internationale des arts et techniques dans la vie moderne opened, Paris (31,000,000 visitors).

Secret missile base was built for construction of experimental liquid-fueled rockets, Peenemünde, Germany.

Premiers: Production of florescent lighting (see p. 80); regular television broadcasting, U.K.; flame-resistant Pyrex, a further enhancement of the product, by Corning, U.S., *LIFE* magazine, U.S.; paperback book not intended for rebinding, by Penguin, U.K.; practical helicopter ("Fa-61"), by Heinrich Foche, Germany; Betty Crocker, fictional typical American housewife, by General Foods, U.S.; supermarket carts, Oklahoma, U.S.; Minox microfilm camera, by Walter Zapp, Latvia; concealed headlamp on an auto, the Cord "812," U.S.; turn-type antenna for television reception, by George Harold Brown; elementary form of digital computer using electromagnetic relays, by Konrad Zuse; isolation of DNA in a pure state, by Andrei Kikolaevitch Belozersky; artificial heart used during cardiac surgery, by Alexis Carrel with Charles Lindbergh.

A battery-operated transistor radio was produced by Braun, Germany.

Boulder Dam (later Hoover Dam) was completed, creating Lake Mead, the world's largest reservoir, U.S.

Oparin's *The Origin of Life on Earth* proposed that life arose randomly in the oceans, Russia.

Louis Blériot, Federico García Lorca, Maxim Gorky, Luigi Pirandello, and Ottorino Respighi died.

"American Modern" dinnerware by Russel Wright

Russel Wright (1904–76) studied in Cincinnati, and New York where he set up a studio. Highly nationalistic, he began to lecture widely and railed against what he considered to be America's lingering feelings of aesthetic inferiority. In his own work, he combined functionalist, Art Déco, and Mission styles in an odd balance, and his wife Mary, a shrewd promoter, made a major contribution to his efforts. In 1935, Wright formed a partnership in New York with businessman Irving Richards to produce products for the home. Richards persuaded Wright to follow the success of his "Modern Living" furniture suite with a dinnerware line. Named "American Modern," the simple service in organic, soft, rimless shapes was glazed in unusual, mottled colors. Almost without precedence, each color was designed to harmonize with all others; customers were encouraged to arrange their own mixed-color table settings. Wright believed that food should be placed on plates in colors most flattering to it. The design of the multi-piece service was completed by 1937. But, due to the unusual shapes, finding a manufacturer was difficult, and, thus, shipment to stores was not made until 1939. Eventually supplemented by glassware, table linens, and flatware, production continued until 1959, by which time 80,000,000 pieces of "American Modern" had been made.

Date: 1937. Material: Glazed earthenware. Manufacturer: Steubenville Pottery Co., East Liverpool, OH, U.S. Photograph: Courtesy the Montreal Museum of Decorative Arts, gift of David A. Hanks (pitcher, D82.105.1), The Liliane and David M. Steward Collection (pepper and salt shakers, cup and saucer, D87.170.2, D88.180.7,8,10a-b). Photo: Giles Rivest, Montreal.

Events of 1937:

Exposition internationale des arts et techniques dans la vie modernes opened, Paris (31,000,000 visitors).

Premiers: A toothbrush with an angled handle like a dental mirror, patented by dentist Fuller, made by Squibb, U.S.; polyurethane; yellow-fever vaccine, by South African microbiologist Max Theiler; racing-dog symbol for the Greyhound Bus Company and ferry boat *Princess Ann*, to influence subsequent steamship design, by Raymond Loewy, U.S.; antihistamine, by pharmacist Daniele Bovet, Pasteur Institute, Paris; xerography, a photocopying method, by law student Chester Carlson, U.S.; electronic computer, begun by John V. Atanasoff; drive-in bank, Los Angeles; franchise restaurant (Howard Johnson); Buick "Y" Job auto, by Harley Earl, U.S.; working jet engine, Frank Whittle, U.K.; patent of Nylon, by du Pont, U.S.

Film: Dieterle's *The Life of Emile Zola*, Wellman's *A Star is Born*, Vidor's *Stella Dallas*, Disney's *Snow White and the Seven Dwarfs*.

Literature: Hemingway's *To Have and Have Not* and Malraux's *Man's Hope*.

The Duke of Windsor and Wallace Simpson were married, France, and King George VI was crowned, U.K.

Tallest dinosaur to this time, Iguanodont, was found by Dr. Barnum Brown, Wyoming, U.S.

Italy banned racially mixed marriages in North Africa.

"Glass Age" exhibition on The Glass Train by Kenneth Cheeseman for Pilkington Brothers traveled throughout the U.K.

Hindenburg dirigible "LZ 129" exploded, New Jersey, U.S. Japan invaded China.

Jean Harlow, Bessie Smith, Guglielmo Marconi, George Gershwin, and Maurice Ravel died.

Spaniard Antonio Bonet (1913–89) and Argentines Juan Kurchan (1913–75) and Jorge Ferrari-Hardoy (1914–77) met while working in the architecture office of Le Corbusier in Paris. Subsequently, Kurchan and Hardoy, with Bonet, returned to Argentina where they created the "B.K.E." chair. It has also been called by other names including the "A" for the two overlapping metal sections, the "Butterfly", and the "Sling." The highly simple standard form had roots in Britain where, in 1877, Joseph Beverley Fenby patented a folding wooden chair, and an American version of the Fenby chair has been continuously produced since 1895. To historian Karl Mang, the chair symbolizes "the 'design that could have been' in Spain if the Rationalist triumphs had continued" into Generalissimo Franco's repressive regime (see p. 188). Soon after 1940, when the chair was awarded a prize in Argentina, New York importer Clifford Pascoe put it into production in America. In about 1946, furniture manufacturer Hans Knoll acquired the rights to make the chair. Due to a proliferation of copies by others in America and Europe, Knoll sued for copyright infringement in 1950 but lost due to the chair's likeness to Fenby's. In the 1950s alone, about 5,000,000 copies were made in steel tubing or solid-steel rods and with cloth or leather slings.

Date: 1938. Materials: Bent steel rods, canvas (example shown here) or leather. Manufacturers: Knoll Associates, East Greenville, PA, U.S.; Gold Metal GmbH, Stöhr (Besingheim), Germany (example shown here); others. Photograph: Courtesy Vitra Design Museum, Weil am Rhein, Germany.

Events of 1938:

Hungarians László and Georg Biró patented a ballpoint pen, France; the ink was perfected later by Frank Seech, California.

The streamline "S1" locomotive was designed for the Pennsylvania Railroad by Raymond Loewy, U.S.

Color television was first shown, by John Logie Baird, U.K.

Premiers: Fiberglass, by Owens-Corning; "Z" binary calculating machine, by Konrad Zuse; nylon-based product (Dr. West brand toothbrush), U.S.; artificial hip replacement, by Dr. Philip Wiles, U.K.; radio altimeter; "Superman" comic strip, U.S.; domestic steam iron; pressurized cabin in a passenger airplane; ECT or "shock" therapy, by Drs. Ugo Cerlutti and Lucio Bini, Italy; Hewlett-Packard; Volkswagen "Beetle," by Ferdinand Porsche, Germany.

Teflon resistant coating was discovered by Roy Plunkett, U.S., first sold as a coated fry pan in 1956, Nice, France.

Film: Howard's *Pygmalion*, Eisenstein's *Alexander Nevski*, Hitchcock's *The Lady Vanishe*s.

Literature: du Maurier's *Rebecca*, Sartre's *Nausea*, Nabokov's *Invitation to a Beheading*.

Industrially emitted carbon dioxide in the Earth's atmosphere was determined by G.S. Callendar.

Liquid-fueled rocket, under Wernher von Braun's direction, successfully traveled 18 km (11 miles), Germany.

"Crystal Night" rampage by young Nazis destroyed Jewish stores, homes, and synagogues, Berlin, and the Nazis entered Austria.

The Pope praised Mussolini and rebuked Hitler, Vatican.

Nuclear fusion was discovered by Otto Hahn and Fritz Strassman, Berlin.

Aramco found the first commercial oil in Saudi Arabia.

Goggles by Samuel E. Bouchard

Certain anonymous, or presumably anonymous, designs abound. When they are good, when they work well, they assume roles as indispensible servants with essentially endless lives. Consider, for example, the safety pin of 1849, the paper clip by a Norwegian scientist but patented by an American in 1900, the practical version of the ballpoint pen of 1938 (see p. 120), and the computer mouse of 1967 (see p. 180). Thanks to the establishment of the U.S. Patent Office in 1790 and recent assiduous research by historians in Europe and elsewhere, the identities of the inventors of certain products have surfaced. Even though little is known about the appearance of eyeglasses in 13th-century Italy, they have had a 700-year history of innovation which continued with the establishment of an optical instrument company in America in 1853 by German immigrants John Jacob Bausch and Henry Lomb. In the early 1930s, the firm was commissioned by the U.S. Army Air Corps to create a material for sunglasses. A glass was developed that screened out 70–90% of visible light and banned 95% of ultraviolet rays, thus the trade name Rayban. Credit has been assigned to Brian O'Brien and Franklyn C. Hutchings. The lenses were fitted to Samuel E. Bouchard's goggles that he patented in 1939. The sunglasses, virtually unchanged to today and certainly an anonymous design, became popular with World War II pilots and ever since with the general public.

Date: 1939 (frame). Materials: Tinted optical glass, plastic material (example shown here of 1975). Manufacturer: Bausch & Lomb, Rochester, NY, U.S. Photograph: © Gregory S. Krum, New York.

Events of 1939:

New York World's Fair ("Building the World of Tomorrow") opened, New York City (45,000,000 visitors to 1940).

Premiers: The pressure cooker, by National Presto Industries, U.S.; operational prototype of electronic computer, by John V. Atanasoff; "2 CV" (Deux Chevaux) prototypes, by Citroën (production from 1948); polyethylene, by ICI, U.K.; hand-held slicing knife, U.S.; jet engine in a flying airplane ("He 178"), by Pabst von Ohain, Germany; precooked frozen food, U.S.; complex-number calculator, by Bell Laboratories, U.S.; flight of what was to become the standard helicopter, by Igor Sikorsky, U.S.; regular commercial transatlantic flights, by Pan American World Airways; DDT insecticide, by chemist Paul Müller, Switzerland; vulcanized rubber, by Charles Goodyear, U.S.; automatic dishwasher; air-conditioned auto, in the Packard; transatlantic mail service, U.S.; "Batman" comics, by graphic artist Bob Kane.

Film: *Gone with the Wind* and *The Wizard of Oz*.

Arabia firm became Europe's largest manufacturer of porcelain, Denmark.

The Germans re-established legally restricted Jewish ghettos, first in Lódź, Poland.

Albert Einstein's letter to President Roosevelt led to the U.S.'s developing an atomic bomb. (The letter was dated 2 August.)

Soviet and German forces began occupying Poland, precipitating World War II (1 September). To thwart Germany from invading Russia, Stalin assaulted Finland for a takeover and as a buffer.

With the Civil War over, Generalissimo Franco assumed power, Spain.

Tokyo Shibaura Electric Co. (Toshiba) was founded, Japan.

Douglas Fairbanks Sr., Anton Fokker, Sigmund Freud, and William Butler Yeats died.

**Minnesota Mining and Manufacturing Company
(3M) in Minneapolis was essentially a sandpaper
company until 1925 when a young laboratory
assistant there, Richard G. Drew, invented
masking tape. The company was transformed into
a successful pressure-sensitive tape manufacturer.
Five years later, 3M responded to a St. Paul
company's need to insulate hundreds of railroad
cars. Drew coated glue onto a new transparent
film, cellophane, that another 3M researcher had
shown to him. After extensive experimentation,
cellophane adhesive tape was invented. It was
later named "Scotch" tape. Because the end of the
tape stuck back onto the roll after a piece was
torn off, a dispenser was necessary for use in
offices and homes. 3M employee John Borden,
developed a metal dispenser, but it was awkward.
The design firm, Barnes and Reinecke of Chicago,
was commissioned to design a new version. On
the patent drawing by Jean Otis Reinecke, a
principal in the studio, a screw bolt appeared
above the front foot, but, due to the nature of
plastics, the two sides were made to snap
together, eliminating the necessity of a bolt
connector. The streamline design of the dispenser
was far more restrained than those for his other
clients, like the rocketship-inspired ice crusher
of 1939 for appliance manufacturer Dazey.
Reinecke was an enthusiastic advocate of plastics
and felt it was unjust to consider them as
imitations of something else.**

Date: 1940. Material: Plastic. Manufacturer: 3M, Minneapolis,
MN, U.S. Photograph: © Gregory S. Krum, New York.
Collection JoAnn Klein.

Events of 1940:

Italy declared war on the Allies. Norway, Denmark, the
Netherlands, Belgium, Luxembourg, and France were invaded
by the Germans who retaliated against Britain by bombing
population centers and destroying Coventry Cathedral.

The Japanese built the fast, agile, bomb-equipped "Zero"
airplane, named for the zero-year, 1940 (2,600 years since
Japan's legendary founding).

Charles and Ray Eames and Eero Saarinen won the first
prize in the "Organic Design in Home Furnishings" competition of The Museum of Modern Art, New York City.

The "Jeep," designed by Karl K. Probst, was first built by
American Bantam, U.S.

Premiers: The automatic gearbox for autos, by General
Motors, U.S.; experimental broadcast of color television
(developed by Peter Carl Goldmark), by CBS from the Chrysler
Building, New York; M&M candy; British Overseas Airways
(BOAC, today British Airways); U.S. social-security check
($22.54), first paid to May Fuller, Vermont, U.S.; employment
of freeze-drying for food, which had earlier been used for med-
icines, U.S.; Bugs Bunny animated by Tex Avery, for Warner
Bros. Studio; "freeway" road to connect the city to the suburbs,
Los Angeles to Pasadena, California.

Film: *My Little Chickadee* with Mae West and W.C. Fields
and *The Great Dictator* with Charlie Chaplin.

Literature: Hemingway's *For Whom the Bell Tolls* and
e.e. cummings's *50 Poems*.

American scientists discovered plutonium, the element that
would fuel the first successful nuclear weapon.

"Nylon Day" (15 May) signaled the first retail sale of
women's hosiery knitted with the new du Pont fiber, U.S.

Peter Behrens, Leon Trotsky, and F. Scott Fitzgerald died.

"Chemex" coffeemaker by Peter Schlumbohm

Born in Kiel, Germany, of wealthy parents, Peter Schlumbohm (1896–1962) received a doctoral degree in physical chemistry in Berlin in 1926. After several trips to the U.S., he settled there in 1936 to benefit from its inventor-friendly patent laws. His most successful product, the Chemex coffeemaker of 1939, was patented in 1941. The apparatus incorporated items found in every chemistry laboratory: the top of a glass funnel, the bottom of a glass Erlenmeyer flask, and a paper filter. Schlumbohm required a wartime approval from the government to have the heat-proof-glass (Pyrex) body made for him by Corning Glass. On the first versions, a square trough curved outward from the top to the waist; the handle was in wood or cork. On the next stage, the square trough extended up to the top edge, creating a pouring spout and allowing steam to escape alongside the paper filter. Finally the trough was rounded, as on today's model. A measuring mark, or button, indicates the useful volume of liquid (one half of the bottom). The pot was assembled by Schlumbohm's own female workers, and he wrote his own advertising, paid his expenses immediately, and worked only 1 or 2 hours a day. At first he met resistance from diffident store buyers who found the pot strange, but, with yearly sales reaching $200,000 by 1945, the Chemex coffeepot made Schlumbohm rich.

Date: 1939-41. Materials: Pyrex (borosilicate) glass, wood, leather, paper (example shown here of the 1940s). Manufacturer: Chemex, New York, U.S. Photograph: © Gregory S. Krum, New York.

Events of 1941:

Without warning, the Japanese attacked an American naval base (7 December), Pearl Harbor, Hawaii, and other U.S.- and British-controlled sites. Germany and Italy declared war on the U.S. German General Rommel had arrived in Tripoli for the North African defense.

Show-business personalities entertained American troops in the U.S. and abroad with the formation of the USO.

Industrial designers Charles and Ray Eames set up a workshop in their apartment, Los Angeles, U.S.

Premiers: An insect spray in aerosol-canister form ("bug bomb"), by L.D. Goodhue and W.N. Sullivan, U.S.; Cheerios dry cereal, U.S.; silver-zinc battery; non-stretching, non-wrinkling, non-fading, and insect-resistant polyester fiber, by John Rex Whinfield of ICI, U.K.; Mount Rushmore National Park (with giant presidential heads by Gutzon Borglum), Idaho, U.S.; aberration-correcting Maxutov telescope, Moscow; industrial production of polyurethane (invented in late 1930s by Otto Bayer), Germany; approval of atomic-bomb construction, by President Roosevelt, U.S.

Film: *Citizen Kane*, written, produced, directed and acted in by Orson Welles, and Houston's *The Maltese Falcon*.

Literature: Hemingway's *The Last Tycoon*, Auden's *The Quest*, Nazi refugee and psychologist Fromm's *Escape from Freedom*, Ehrenburg's *The Storm*.

Music: Britten's "Nighthawks" and the popular song "Chattanooga Choo-Choo."

Grand Coulee Dam became the first man-made structure to exceed the volume of the Great Pyramid of Cheops, U.S.

Henri Bergson, James Joyce, and Virginia Woolf died.

Bruno Mathsson (1907–88) was a fifth-generation cabinetmaker. Working with wood from childhood in his father's workshop in Vüarnomo, Sweden, Mathsson acquired an appreciation and knowledge of the material that was so readily available in Scandinavia. Of the new age, Mathsson hungrily read foreign publications on design and steeped himself in ideas of the nascent Modern Movement. His first practical experiment, the "Grasshopper" chair of 1931, was not successful but whetted his appetite for further investigation into bending wood. Initial international acceptance and sales of his furniture came in 1937 when Mathsson's new, refined chairs, including the "Eva" of 1934, were shown at the international exposition in Paris. Always returning to his home base, Mathsson traveled widely and made a trip to the U.S. in the 1940s where he met some of its design giants, a visit that precipitated the development of his glass houses. However, the "Miranda" chair with its "Mifot" footstool (not shown) is the design for which he may be best known. A paper yarn, employed even during the war, is still the standard upholstery material, but customers can provide their own fabric. Mathsson has been favorably compared to better-known Modernists like Alvar Aalto and Marcel Breuer. But Mathsson's sexy forms—all with female names—are members of an exclusive sorority.

Date: 1942. Materials: Beechwood, round-woven paper yarn or hemp. Manufacturer: Bruno Mathsson International AB, Vüarnomo, Sweden. Photograph: Courtesy Bruno Mathsson International AB.

Events of 1942:

South Pacific conflict raged with the Allied forces against Japan and included the Battle of the Coral Sea, Battle of Midway, closing of the Burma Road, and the Battle of Guadalcanal. Fierce fighting and bombing continued in Europe and Russia. Under the Vichy regime in France, anti-Semitic laws were passed, dissidents were jailed, and "Liberty, Equality, Fraternity" was changed to "Work, Family, Fatherland."

Premiers: The bazooka rocket gun; napalm liquid explosive; Raisin Bran dry cereal; K-ration food for soldiers, U.S.; ABC (Atanasoff-Berry computer, considered the prototype of all subsequent electronic computers), by John V. Atanasoff and Clifford Berry, U.S.

U.S. women began serving (eventually reaching 350,000) in the armed forces, but not in combat.

Due to food shortages, "victory gardens" were being planted by civilians; U.K. and U.S. food rationing began with sugar at first, followed by coffee, gasoline, flour, fish, and canned goods.

U.S.'s OSS and U.K.'s SOE joined for wartime sabotage and espionage, with over 11,000 agents.

J. Robert Oppenheimer and Brigadier General Leslie R. Groves became heads of the Manhattan Project for the atomic bomb's construction, U.S.

Reduced Tour de France race route avoided demarcation.

Charles and Ray Eames introduced organic design forms, U.S.; Le Corbusier created the "Modular," France.

Film: Disney's *Bambi* and Lubitsch's *To Be or Not to Be*. Film studio insured actress Betty Grable's legs for $1,000,000 through Lloyd's of London.

Literature: Thurber's *The Secret Life of Walter Mitty*, Camus's first novel *L'éstranger*, Dinesen's *Winter Tales*, Glasgow's *In This Our Life*.

Due to the death and destruction caused by World War II in Britain, the country was in dire straits, especially in urban areas. Clothing, ceramics, furniture, and other goods were being manufactured according to the government's Utility Scheme and were rationed. Timber shortages prevailed, and damage from bombing boosted the demand for furniture, raising prices. The government established a program to produce "Standard Emergency Furniture" and registered the Utility mark "CC41," which had to appear on every new piece of furniture. Furniture designer and maker Gordon Russell (1892–1980) had final approval of designs. The first series was called the Cotswold collection. Since Cotswold had been the 19th-century seat of the Arts and Crafts movement, the name alone might offer some insight into the utilitarian or modern imperative that was combined with a retrograde attitude. The first collection of about 30 pieces arrived in shops early in 1943. But this new furniture was available only to newly-weds or people whose furniture had been destroyed by bombing. Others had to choose from a restricted range or purchase second hand. Regulations controlling the appearance of Utility furniture were rescinded in 1948, though price regulations remained. The Utility program did not finally end until 1953 when there were about 2,500 firms making the tax-free furniture.

Date: 1943 (example shown here). Materials: Oak, Rexine (plastic-coated fabric). Manufacturer: Utility Scheme, London, U.K. Photograph: By kind permission of the Trustees of the Geffrye Museum, London.

Events of 1943:

War continued worldwide with the hopeless Warsaw Ghetto uprising of the Jews; Siege of Leningrad; the British in Tripoli, North Africa, and German General Rommel's retreat; U.S.'s retaking Guadalcanal; Allied invasion of Sicily; end of the Battle of the Atlantic; Italy's declaring war on Germany.

American railroads were seized by the government for troop transportation and railroad workers were paid extra.

IKEA was founded, Sweden.

The patented design of furniture using simple wood and surplus military webbing was designed by Jens Risom for Knoll before he entered the U.S. Army, U.S.

Premiers: Aqua-Lung, by Jacques Cousteau, France; American Broadcasting Co.; Jefferson Memorial, Washington; automatically withheld income tax, U.S.; Polaroid Land camera, by Edwin Land, U.S.

Literature: de Saint-Exupéry's book for children *The Little Prince*, Schwartz's *Genesis*, Maritain's *Christianity and Democracy*, Eliot's *Four Quartets*, war journalist Pyle's *Here Is Your War*, Rand's *The Fountainhead*.

Music: Bartók's "Concerto for Orchestra," Valazquez and Skylar's song "Besame Mucho," Arlen and Mercer's song "One for My Baby."

Piet Mondrian painted *Boogie-Woogie* in New York.

Theater: Brecht's "The Good Woman of Setzuan," Weill and Nash's musical comedy "One Touch of Venus," Rogers and Hammerstein's musical "Oklahoma!" with DeMille's choreography.

Lorenz Hart, Leslie Howard, Beatrix Potter, Sergei Rachmaninov, Chaim Soutine, and Simone Weil died.

Few people have been adept and financially successful at both design and fine art. But Isamu Noguchi (1904–88) was. Yet, he was very poor during the early part of his career. From 1927–29, he was supported by a Guggenheim Fellowship in Paris where he assisted sculptor Contantin Brancusi for a few months. In 1926, Noguchi designed his first utilitarian object, a plastic clock called "Measured Time." Not until 1937 was his first significant utilitarian object realized; Zenith made the helmet-like Bakelite "Radio Nurse," an intercom for parents to hear their infants remotely. Active as a sculptor all the while, Noguchi designed his first table for the home of A. Conger Goodyear, the president of New York's Museum of Modern Art. Noguchi became miffed when interior designer Terence Robsjohn-Gibbings pilfered the idea of the table's free-form base and glass top. One of several Noguchi versions (shown here) was mass produced from 1944. The two parts of the base, essentially identical, are held by a brass pin where they meet. Since the table could be knocked down, shipment by the manufacturer was economical. Noguchi declared: ". . . I wanted something irreducible, an absence of the gimmicky and clever." The sculptural form announced the post-war development of organic design by Noguchi and others (see p. 104).

Date: 1944. Materials: Painted wood or stained walnut, glass. Manufacturers: Herman Miller Furniture Company, Zeeland, MI, U.S. (1944–73, from 1984); others. Photograph: Courtesy Herman Miller Furniture Company.

Events of 1944:

Led by Wernher von Braun, the Germans put the autopiloted "V-2" rocket into service, relentlessly bombing London.

U.S. forces advanced across the Pacific. Allied forces crossed the English Channel and invaded France at Normandy (6 June). The Battle of the Bulge ensued.

Anne Frank and her family were seized and sent by cattle car to Auschwitz where she died in 1945.

London's Council of Industrial Design was founded; the Old Vic Theatre was revived, and actors Richardson and Olivier returned from military service to produce what some consider the birth of the modern English theater, London.

Premiers: The antibiotic Aureomycin, by Benjamin Minge Duggar, U.S.; liquid-crystal clock (ten times more accurate than the pendulum type), at the Greenwich Royal Observatory, U.K.; Automatic Sequence Controlled Calculator (second-generation electronic digital computer), by Howard Aiken and team of engineers at IBM; Chiquita Banana; *La Monde* newspaper, Paris; negro voting, Texas.

Francis Bacon's painting *Three Studies for Figures at the Base of a Crucifixion* drew comparisons to Goya, London.

Film: Olivier's *Henry V* and Hitchcock's *Lifeboat*.

Literature: The play "La Folle de Chaillot" (Madwoman of Chaillot) by former diplomat Giraudoux who died shortly after, Sartre's four-character existentialist play "No Exit," Maugham's *The Razor's Edge*, Lagerkvist's book *The Dwarf*, Camus's book *Caligula*.

Leo Baekeland, Alexis Carrel, Wassily Kandinsky, Edwin Lutyens, Aristide Maillot, Glenn Miller, Piet Mondrian, Edvard Munch, Erwin Rommel, and Antoine de Saint-Exupéry died.

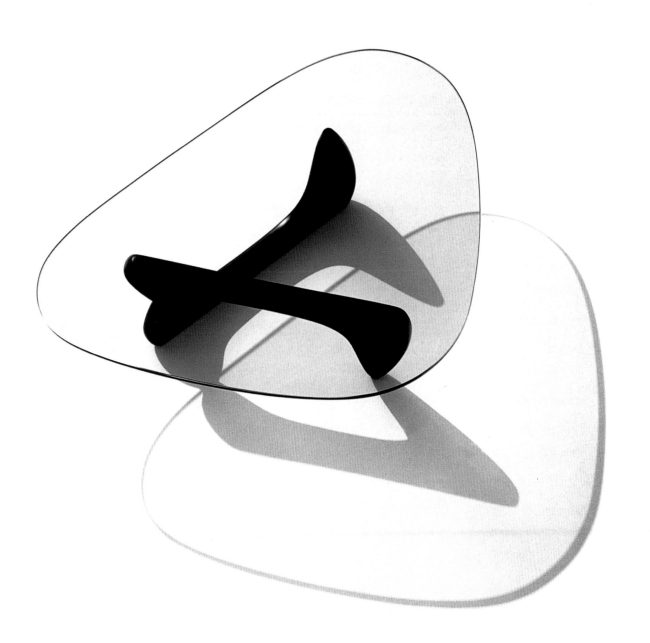

ICI of Britain developed polyethylene, a synthetic polymer, in 1939 and offered it commercially soon thereafter. American inventor and engineer Earl C. Tupper (1907–83), a chemist at du Pont in Massachusetts, discovered a technique for injection molding polyethylene in 1942 and established the Tupper Plastics firm in 1945 for the production of lightweight, unbreakable, crack-resistant, translucent, and hygienic kitchen containers. His first product in "Poly-T," Tupper's name for polyethylene, was a 7-oz. bathroom tumbler; it was followed by a range of 25 different pastel-colored cups, bowls, pitchers, and other items, some frivolous. A later snap-on-lid feature on some containers kept refrigerated foods fresher. From 1946 in the U.S. and 1960 in Britain, Tupper's revolutionary wares were sold by housewives—who were paid commissions—directly to other housewives at special Tupperware parties. The demonstration samples at the parties were not sold, but rather orders were later shipped to customers. Tupper was planning to move his factory to larger quarters when he abruptly changed his mind, sold the firm to the Rexall Drug Company in 1958, and retired to Costa Rica. Tupperware has retained its success through current production of a huge number of products in materials other than polyethylene, designed today by Morrison Cousins and others.

Date: 1945. Material: Molded polyethylene. Manufacturer: Tupper Plastics, Farnumsville, MA, U.S. Photograph: Courtesy Tupperware World Headquarters, Orlando, Florida, U.S.

Events of 1945:

Germany surrendered to the Allies, ending the European portion of World War II (7 May).

Fearing Germany's development of an atomic bomb, an experimental example (attributed to German immigrant Otto Frisch and Dutch physicist Niels Bohr) was detonated at Los Alamos, Nevada, U.S. Subsequently atomic bombs were dropped on Hiroshima (6 August) and Nagasaki (9 August). Japan then surrendered to the Allies, thus ending World War II in the East.

15,900,000 military people died in World War II. Liberated Nazi concentration camps revealed the systematic murder of 6,000,000 Jews and others.

The U.S. captured and shipped to America 100 V-2 rockets, the Peenemünde-base archive, and 115 German scientists, including physicist and SS officer Wernher von Braun who later developed America's rocket program.

Science-fiction writer Arthur C. Clarke proposed the idea (effective from 1965) of communications satellites in space.

Premiers: The fluoridation of public water to prevent tooth decay, Grand Rapids, MI, U.S.; female suffrage, France; zoom camera lens; wax or china-marking pencil; national televising of the Macy's Thanksgiving Day Parade, New York; Hasselblad camera, by Sixten Sason, Sweden; frozen orange juice, U.S.

White Sands proving ground for rocket research was established, New Mexico, U.S.

Literature: Orwell's *Animal Farm*, Waugh's *Brideshead Revisited*, Mitford's *The Pursuit of Love*.

Films: Rossellini's *Open City* and Wilder's *The Lost Weekend*.

Charles de Gaulle was elected president of France.

Franklin Roosevelt died, and Harry Truman became President, U.S. Adolf Hitler committed suicide, Berlin. Béla Bartók, Anne Frank, Käthe Kollwitz, Benito Mussolini, Alla Nazimova, Paul Valéry, and Anton von Webern also died.

From 1936–40, Charles Eames (1907–78) studied and taught at Cranbrook Academy in Michigan, where he began a collaboration with Eero Saarinen, son of the school's director. They came into prominence in 1940 when their chair design received a first prize at the organic-design competition of New York's Museum of Modern Art. Eames who had married Ray Kaiser (1912–88), a Cranbrook student, settled in California. In the early 1940s, the Eameses built a rather dangerous plywood-bending contrivance they called the "Kazam!" machine. It continued earlier attempts to make a plywood chair. Even though the Eameses' machine was not appropriate for mass production, they succeeded in bending plywood into 3-dimensional shapes, not possible by any-one before. Rather than pursuing with the one-piece shells of the late 1930s that split at sharp curves, Charles and Ray in California decided to separate the back from the seat. By 1945, they had developed what has become known as the "Potato Chip" chair, with bent plywood legs in low- and dining-height versions and soon devised a method for mass production. The designation here, "LWC," represents "low wood chair." Rubber disks that separate the curvaceous seat and back from the frame offer a slight squishy movement and thereby a sense of comfort.

Design: 1945–46. Materials: Black-stained plywood, rubber, metal (example shown here of 1946). Manufacturer: Evans Products (1946); Herman Miller Furniture Company, Zeeland, MI, U.S. (1946–53, from 1994). Photograph: Courtesy Montreal Museum of Decorative Arts, gift of Ann Hatfield Rothschild (D81.152.1). Photo: Richard P. Goodbody, New York. By permission Eames Office, www.eamesoffice.com.

Events of 1946:

Nuremberg trials were conducted to punish war criminals, Germany.

"Britain Can Make It" exhibition opened, London.

Charles Eames's exhibition, the first one-person show of a designer at The Museum of Modern Art, was organized by Eliot Noyes, New York.

Premiers: The bikini swimsuit; radial auto tire ("Michelin X"), France; all-purpose stored-program electronic computer, the ENIAC (Electronic Numerical Integrator and Computer), by J. Presper Eckert Jr. and John Mauchly and built at the University of Pennsylvania, U.S.; "Vespa" scooter, designed by Corradino D'Ascanio and produced by Piaggio, Italy; portable electric drill ($16.95), by Black & Decker, U.S.; Tide clothes-washing detergent; Timex watch; prototype of the "4 CV" auto, by Renault, France; microwave oven, by Percy Le Baron Spencer (sold from 1947), U.S.; availability of the food freezer for domestic use, U.S.; female suffrage, Japan; Slinky toy; Cannes film festival, France.

Film: De Sica's *Shoeshine*, Capra's *It's a Wonderful Life*, Cocteau's *Beauty and the Beast*.

Erick Hodgins coined the phrase "dream house" in his novel, *Mr. Blandings Builds His Dream House* (filmed in 1948 with Cary Grant), U.S.

First meeting of the United Nations was held.

Ruth Benedict's *The Chrysanthemum and the Sword*, financed by the U.S. government, attempted to explain the Japanese culture to Westerners.

The radioactive carbon-14 method of dating ancient objects was introduced by Willard Frank Libby, U.S.

First Soviet nuclear reactor went into use under the direction of Vasilevich Kurchatov.

John Logie Baird, Gertrude Stein, and H.G. Wells died.

Rudolph Wurlitzer (b.1829) was born into a family of musical-instrument makers and vendors in the Saxony area of Germany. At age 24 he settled in Cincinnati, Ohio, and in 1856 founded the Rudolph Wurlitzer Company. At first he imported musical instruments from his family and later began making them himself. He produced his first piano in 1880 and the first coin-operated electric piano, the "Tonophon," in 1896. The organs made in New York State from 1911 for movie-houses and theaters that became popular during the silent-film era were designed by English immigrant Robert Hope-Jones and were hyped as "Mighty Wurlitzers." Wurlitzer's automatic musical soundtrack system for movies was an innovation like his jukebox that was introduced in 1933. Were it not for the window on the first model, the "Debutante," it might have been mistaken for any other large piece of furniture in the Queen Anne Revival cum Neo-Renaissance style of the time. But jukeboxes were almost never installed in homes then. A succession of models to 1939, when production was discontinued due to the war, had gradually been transformed into soft-edged, streamline forms; some wood elements were replaced by plastics. But it is the bombastic "Model 1015" of 1946–47 by Wurlitzer technician Paul Fuller that became the quintessential jukebox. The machine came alive when rising bubbles percolated around the sides and light danced behind garish, rainbow-hued plastic panels. More than 56,000 units were produced, and reproductions are being made today.

Date: 1946–47. Materials: Wood, plywood, metal, plastics, glass (reproduction shown). Manufacturer: Rudolph Wurlitzer Company, Cincinnati, OH, U.S. Photograph: Wurlitzer Jukebox Company.

Events of 1947:

The U.S. Marshall Plan to help rebuild devastated countries was established.

The grandiose coffee machine by Gio Ponti for Pavoni became an organic, sculptural symbol in espresso bars of the 1950s in Italy.

The New Look in women's clothing was launched by fashion designer Christian Dior, Paris.

Le Corbusier built the Unité d'Habitation, Marseille.

Designer-manufacturer cooperative organization Deutsche Werkbund was re-established, Germany.

Premiers: Aluminum foil (a .0007 inch thick sheet) for domestic use, introduced by Reynolds Metals, U.S.; food processor, by Kenneth Wood, U.K.; Coca-Cola soda-fountain dispenser and Studebaker "Champion" auto, both designed by Raymond Loewy, U.S.; basic concept of holography (not practical until the laser's invention in the 1950s), by Hungarian-British physicist Dennis Gabor; tubeless tire, by Goodyear, U.S.; "Howdy Doody" children's television program, on NBC, U.S.; transistor (tiny device like a vacuum tube but requiring much less power), by William Shockley, John Bardeen, and Walter H. Brattain at Bell Laboratories, New Jersey, U.S.; geodesic domes, developed and built by R. Buckminster Fuller, U.S.

Pilot Chuck Yeager broke the speed-of-sound record, U.S.

Two shepherd boys discovered the Dead Sea Scrolls, the earliest-known manuscripts of the Bible, Qumran, Palestine. Archaeologists later uncovered more manuscripts.

Thor Heyerdahl sailed the handmade *Kon-Tiki* boat from Peru to a Pacific island, to prove migration from the East.

Based on dreamy cut-and-pasted hand-painted paper, Matisse published the book *Jazz*, France.

Literature: Mann's *Doktor Faustus*, Calvino's *The Path to the Next Spider*, Quasimodo's *Day After Day*.

Pierre Bonnard, Ettore Bugatti, and Henry Ford died.

The Museum of Modern Art in New York, as part of its "International Competition for Low-Cost Furniture Design" of 1948, granted money to designers, including Charles Eames (1907–78), for them to work closely with manufacturers to apply new technology to furniture production. Eames, his wife Ray (1912–88), and members of their staff partnered with the Engineering Department of the University of California in Los Angeles. The team's self-assigned mandate, like the Saarinen-Eames project for the museum's "Organic Design in Home Furnishings" of 1940, was to create a chair with the least number of parts and a mass-production potential. They used steel, much cheaper than plywood, and sprayed the surface with a synthetic rubber, neoprene, because the metal surface was cold. The chair won a second prize in the museum's competition. However, the molds for stamping, the same as those for forming automobile fenders, were fatigued after three or four very heavy chairs were produced. Fortuitously, by 1950, the Herman Miller Furniture Company could mold colored fiberglass into a 3-dimensional, lightweight shell, like the bucket of the "DAX" chair. A number of leg-base options were offered on the "DAX," including the one shown here, nicknamed the "Eiffel Tower" base.

Date: 1948–50. Materials: Fiberglass, metal wire, rubber. Manufacturer: Herman Miller Furniture Co., Zeeland, MI, U.S. Photograph: Courtesy Montreal Museum of Decorative Arts, The Liliane and David M. Stewart Collection, Canada (D81.100 .1). Photo: Richard P. Goodbody, New York. By permission Eames Office, Venice, CA, www.eamesoffice.com.

Events of 1948:

McDonald brothers, using assembly-line methods, opened their first hamburger restaurant, Pasadena, California, U.S.

Premiers: Honda Motorcycles, Japan; long-playing phonograph record, by Hungarian-American physicist Peter Mark Goldmark, U.S.; production of the Citroën "2 CV" (Deux Cheveux) auto, by Citroën, France; Cisitalia "Type 202" automobile body, by Pininfarina, Italy; British Land Rover auto (for farmers' use after World War II), by Land Rover; Porsche auto, by Ferdinand Porsche, Germany; injection of cortisone for relief of arthritis pain, by Philip Showalter, U.S.; World Health Organization of the United Nations.

Mark I stored-program electronic computer prototype began operation; apocryphally a moth created a malfunction, thus the term "bug" (or computer malfunction), U.K.

Mechanization Takes Command (*La mécanisation au pouvoir*) was published by Sigfried Giedion.

Norbert Wiener published his theory of cybernetics, a landmark study of how humans handle information from electronic devices, U.S.

Big Bang theory of the Universe's origin was developed by George Gamow, Ralph Alpher, and Robert Herman, U.S.

Demonstrated in 1947, Polaroid "Model 95" instant-picture camera and film, by Edwin Herbert Land, was put on the market, U.S.

Film: Olivier's *Hamlet*, Huston's *The Treasure of the Sierra Madre*, Litvak's *The Snake Pit* with de Haviland.

Studying cockleburs attached to his clothing, Georges de Mestral invented Velcro (patented 1955), Switzerland.

Sergei Eisenstein, D.W. Griffith, Louis Lumière, and Orville Wright died.

Through the considerable contributions of Scottish inventor John Logie Baird (1888–1946), Britain had a significant presence during the pioneering days of television technology. In 1923, Baird produced a mechanical-scanning TV device and two years later successfully transmitted the first TV image of a recognizable human face. By 1926, he was able to create TV images of moving objects. Then in 1928, he broadcasted a TV signal from England to America. Even though Vladimir Zworykin had applied for a color-television patent in the U.S. in 1926, the year of Baird's transatlantic TV broadcast, it was Baird who first showed color television in 1938. But even black-and-white TV sets were not in general use until after the war, when the best-known model in Britain became the Bush set with a 12- or 22-in. picture tube. The shape of the Bakelite case of the TV and other Bush electronics, like its equally popular "DAC90" radio, continued a predilection for the streamline forms of the 1930s. The use of the plastic material Bakelite, patented by Leo Baekeland in the U.S. in 1907, remained active into the 1950s. The dark brown color was the easiest to produce, but Americans were able to make Bakelite objects in vibrant colors. When 405-line TV tubes were upgraded to 625-line versions in 1964, the popular Bush TV became obsolete. A story tells that "Bush TV 22" sets, shipped to Japan, were turned into fish tanks.

Date: 1948–49. Materials: Bakelite, metal, glass. Manufacturer: Bush, England. Photograph: Courtesy Die Neue Sammlung, Munich, Germany, inv. no. 1414/82.

Events of 1949:

The first industrial designer to receive such prominence, Raymond Loewy appeared on the cover of *Time* magazine.

Soviet Socialist Realism was imposed throughout Eastern Europe.

Premiers: Silly Putty child's toy, U.S.; daytime-television serial drama (soap opera), on NBC, U.S.; situation comedy, on CBS, U.S.; non-stop around-the-world airplane flight; Radio Free Europe broadcasts; "Lego" children's construction blocks, by Ole Kirk Christiansen, Denmark; Soviet atomic-bomb explosion; NATO (North Atlantic Treaty Organization) for a common defense of Western nations against the U.S.S.R.; prepared cake mixes in a box, by General Mills and by Pillsbury, U.S.; Emmy awards, by American Television Academy of Arts and Sciences; operation of the ENIAC automatic binary computer, U.S.

American jeweler Harry Winston bought the 45.5-carat Hope diamond from the estate of Evalyn Walsh McLean, U.S.

Literature: Orwell's futuristic novel *Nineteen Eighty-Four*, Moravia's *Conjugal Love*, Mahfour's *The Beginning and the End*, Weil's *The Need for Roots*.

Mao Zedong proclaimed the People's Republic of China as a communist state and took over as principal ruler.

Germany was split into eastern and western sections.

Fred Lawrence Whipple suggested that comets are "dirty snowballs," made of ice or ammonia-ice and rock dust, U.S.

Cape Canaveral, Florida, was established as a U.S. rocket-testing facility.

26 counties of Ireland, extricated from 800 years of British domination, became the Irish Free State.

James Ensor, Maurice Maeterlinck, José Clemente Orozco, "Bojangles" Robinson, and Sigrid Undset died.

"Antelope" armchair by Ernest Race

English furniture and industrial designer Ernest Race (1913–64) studied interior design at an architecture school that was part of London University and, in the late 1930s, weaving in India. From 1935, he was a model maker and then designed lighting for Troughton and Young. In 1945 at war's end, Race and J.W. Noel Jordan founded Race Furniture Ltd and began using airplane scrap as raw material. Race, who also worked with bent plywood and steel rods, produced the 1949–50 "Springbok" and "Antelope" chairs. They were chosen by the organizers of the Festival of Britain of 1951 to furnish its buildings and grounds on London's South Bank. Zealously promoting British achievements in art, industrial design, science, and technology, the exposition offered visitors a diversion from the austerity they had had to endure during the war years when government-sponsored Utility furniture had been their only choice (see p. 98). One of the buildings, the Dome of Discovery, featured snowflake and atomic patterns. Race's lighthearted chairs furthered the aspirations of the exposition and the new attitude toward science by including, possibly lamely, the idea of an atom's structure in the form of balls at the end of the chair legs. The structure of the chair was a modified version of the steel-rod skeleton inside his model "DA2" upholstered club chair.

Date: 1949–50. Materials: Enamel-painted steel rod, plywood. Manufacturer: Race Furniture Ltd, London, U.K.; others. Photograph: Courtesy Montreal Museum of Decorative Arts, gift of Barbara Jacobson (D88.195.1). Photo: Richard P. Goodbody, New York.

Events of 1950:

Braun produced its first electric razor "S 50" (designed in 1938) to broaden its post-war range of appliances, Germany.

Lake Shore Drive Apartments were designed by Ludwig Mies van der Rohe, Chicago (to 1952). The chapel at Ronchamp was built by Le Corbusier, France (to 1954).

Premiers: "Good Design" competition-exhibition, by and at The Museum of Modern Art, New York; commercial photocopier (xerography), by Xerox, U.S.; Club Méditerranée, France; prototype of the credit card, by Diner's Club (available to a few hundred restaurants, but the petrol-station credit card already existed), U.S; commercial kiss-proof lipstick ($1), by Hazel Bishop, U.S.; general, commercial color television broadcasting, U.S.; national television set in Japan, designed and made by Matsushita Denki; Columbia Broadcasting System (CBS) trademark/logo "eye," by William Golden, U.S.; "Corvette" auto (produced from 1953), by Chevrolet; Studebaker "Starliner" auto, designed by Raymond Loewy; "Little Folks" (later "Peanuts") comic strip, by Charles Schulz, U.S.; cow embryo transplant; "Lettera 22" typewriter, by Marcello Nizzoli for Olivetti, Italy.

"Christmas Hurricane," because it struck at Christmas-time, wreaked havoc in Western Europe.

Korean War began when North Korean troops invaded South Korea, precipitating U.N. intervention.

A computer (the ENIAC) was first used to make 24-hour weather forecasts.

Pollock's and de Kooning's paintings signaled a new, radical movement, to become known as Abstract Expressionism.

Wilder's film *Sunset Boulevard* featured Gloria Swanson.

Literature: Neruda's *Canto General*, Pavese's *The Moon and the Bonfire*, Onetti's *A Brief Life*.

Theater: Anouilh's "The Rehearsal" and Williams's "Rose Tattoo."

"Superleggera" side chair by Gio Ponti

Italy's most accomplished and revered 20th-century architect and designer, Gio Ponti (1891–1979), studied at the Politecnico in Milan from 1918–21 and at first collaborated with others in an architectural office. His credentials became extensive and impressive. The founder in 1928 of the architecture-design journal *Domus*, Ponti's design work ranged from furniture, ceramics, fabrics, enamels, mosaics, and lighting to dinnerware, door handles, sanitary equipment, and even automobile bodies. His giant espresso machine of 1948–49 by Pavoni owed a great deal to the streamline form in America at the time. Ponti was modern in essence but drew on traditional motifs and imagery, manifesting the eclecticism of the early years of fascism. He eventually abandoned neoclassicism, but his continuing interest in reinterpreting traditional forms is illustrated by the "Superleggera," a model based on the classic, provincial Chiavari-chair type. Production required both handcrafting methods and industrial techniques to create the chair's subtle details. For example, the triangular, super-thin elements are slightly rounded. "If you go to Cassina [the manufacturer] you'll be amazed to see these chairs fly through the air, hit the ground, and bounce back still in one piece. . . you're continuously having to dodge flying chairs," Ponti wrote in *Domus* (March 1952).

Date: 1951. Materials: Ash or stained walnut, woven fiber. Manufacturer: Cassina, Milan, Italy (from 1957). Photograph: Courtesy Cassina U.S.

Events of 1951:

Festival of Britain exhibition celebrated the centenary of the Great Exposition of 1851, London.

The first color televisions were sold, U.S.

Industrial designer Raymond Loewy's book *Never Leave Well Enough Alone* secured his reputation and success.

Premiers: Crumple zones in auto bodies, by Bella Barenyi, Germany; Buick "Le Sabre" prototype auto body, by Harley Earl, U.S.; Aspen International Design Conference, Colorado, U.S.; "super glue" (ethylcyanhoacrylate film), by Drs. Harry Coover and Fred Joyner of Eastman Kodak; Council for Design, Darmstadt, Germany; Institut d'Esthétique Industrielle, by Jacques Viénot, Paris; public exhibition of 3-D motion pictures, requiring special polarizing paper-frame eyeglasses, U.S.; regular transcontinental television broadcasting, U.S.; power steering in autos, by Chrysler, U.S.; "zebra" marking at street intersections requiring cars to halt for pedestrians, U.K.

UNIVAC I (Universal Automatic Computer) was the first commercial electronic computer and the first to store data on a magnetic tape, by J. Presper Eckert Jr. and John Mauchly, U.S.

Heating to make a hydrogen bomb from an atomic bomb explosion was achieved by Stanislaw Ulam and Edward Teller.

Tibet became a "national autonomous region" of China, or essentially dissolved as a nation.

Film: Minnelli's *An American in Paris*, Huston's *African Queen*, Bresson's *Diary of a Country Priest*.

Music: Rogers and Hammerstein's musical "The King and I," Britten's "Billy Budd," Menotti's "Amah and the Night Visitors," Gordon's song "Unforgettable."

Literature: Camus's *The Rebel*, Selinko's *Desirée*, Wouk's *The Caine Mutiny*.

André Gide, Sigmund Romberg, and Arnold Schönberg died.

Intended as a see-through object that would float in space, this chair was designed by Arieto "Harry" Bertoia (1915–78). On invitation from furniture manufacturers Hans and Florence Knoll, Bertoia left California, settled in Pennsylvania, and spent two years developing a chair range which included two sizes of diamond-shaped chairs, a high-back lounge chair, an ottoman, a side chair, and a bench. All seating in the series required special jigs and forms specially made to accommodate the production of the essentially handmade chairs. The larger version of the diamond-shaped chair (shown here) could be gently rocked back and forth due to four rubber shocks that suspended the seat from the frame. The bucket seat is more often than not covered with foam-padded fabric or vinyl upholstery to provide protection from the rather brutal wire grid. Unfortunately, upholstery obscures the chair's main feature—its delicate transparency. The leg frame of the smaller version of this chair (without rubber shocks) is bolted directly to the basket seat. Bertoia designed a small number of pieces for Knoll; his main interests centered less on furniture than sculpture, particularly architectural-site installations. Examples of his moving wire sculpture could be seen in Knoll room settings in its literature of the 1950s and '60s.

Date: 1953. Materials: Chromium-plated steel wire, rubber. Manufacturer: Knoll Associates, East Greenville, PA, U.S. Photograph: Vitra Design Museum, Weil am Rhein, Germany.

Events of 1952:

Israel and Germany agreed on restitution for damages perpetrated against Jews by the Nazis.

Christian Dior became the most influential couturier.

The last trams were retired, London.

Challenging conservative views on sex, George Jorgensen, a man, surgically became Christine Jorgensen, a woman, Copenhagen-New York.

Premiers: The semi-domestic microwave oven, by the Tappan Company; appearance of "do-it-yourself" in print; British atomic bomb; American hydrogen bomb; availability of the contraceptive pill (phosphorated hesperidin), U.S.; Streptomycin, by Selman A. Waksman; transistorized radio, by Sony; sugar-free soft drink ("No-cal" ginger ale).

Film: *Les vacances de Monsieur Hulot* with Jacques Tati, *Limelight* with Chaplin, *Othello* with Orson Welles, *Moulin Rouge* (de Toulouse-Lautrec biography) with José Ferrer, Donen's *Singin' in the Rain* with Gene Kelly.

Art: Pollock's *Number 12,* and Dufy's *The Pink Violin*; Epstein sculpted *Madonna and Child,* Cavendish Square, London.

Literature: Hemingway's *The Old Man and the Sea*, Waugh's *Men at Arms*, Frank's *Links wo das Herz ist*, Ferber's *Giant*, Steinbeck's *East of Eden*, Revised Standard Version of the Bible in English for Protestants (touted as "the greatest Bible news in 341 years").

Theater: Christie's "The Mousetrap," continuing for 47 years, becoming the longest-running play ever by 1999; Shaw's "Don Juan in Hell"; Anouilh's "The Waltz of the Toreadors"; Ionesco's "The Chairs."

Alain Bombard crossed the Atlantic Ocean in the inflated canoe "Zodiac M3."

U.K. King George VI and Ferenc Molnár died.

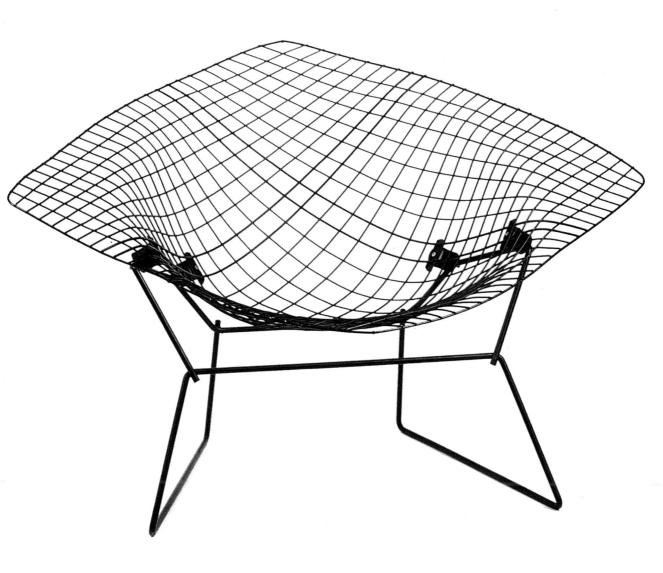

A student of silversmithing in Paris, Serge Mouille (1922–88) for a time worked for the metalworker and sculptor, Gilbert Lacroix. At the end of the war, in 1945, Mouille set up his own studio in Paris and taught at the École des Arts Appliqués. In 1953, he began designing lighting fixtures, at first for Jacques Adnet, director of design of the Compagnie des Arts Français, an artists' cooperative. Mouille has become best known for these lamps with their black *œils* (eyes), as he called them. The cannisters that house the bulb are reminiscent of the "eyes" that extended from the alien craft in the American film *The War of the Worlds*, made the same year as the fixture here. The rays which emanated from the menacing eyes in the movie vaporized everything in sight. Mouille's more friendly eyes—attached to delicate tubular extensions—were thin-metal cups that had been squeezed into the shape of a human-eye opening. For a wide range of applications, Mouille made wall, table, floor, and ceiling versions with single or multiple cups. The lamps of the *œil* series are far more appreciated today than in the 1950s. Mouille completed a number of commissions for schools, even a cathedral in Tunisia. Interested in encouraging young lighting designers, he established the SCM (Société de Création de Modèles) in 1961 and remained active in silversmithing and other metalworking.

Date: 1953. Material: Steel. Manufacturer: Serge Mouille, France. Photograph: © Gregory S. Krum, New York. Collection Galerie de Beyrie, New York, U.S.

Events of 1953:

The Rosenbergs, convicted in 1951 of espionage (possibly wrongly and against public outcry), were electrocuted as the first and only non-wartime civilians killed for spying in the U.S.

Designing for People, based on human factors or ergonomics, was published by designer Henry Dreyfuss, U.S.

Industrial designer Raymond Loewy appeared on the cover of *Der Spiegel*.

Premiers: Commercially available Boeing "707" airplane; "Cristal" ballpoint pen, by Bic; Hochschule für Gestaltung (architecture and industrial-design school based on Bauhaus principles, to 1965), organized by Max Bill and others, Ulm, Germany; Compagnie de l'Esthétique Industrielle design office of Raymond Loewy, Paris; research office Technès, by Jacques Viénot, Paris; initial trial (Calais-Dover) of amphibious hovercraft, pioneered by Sir Christopher Cockrell; Japan Airlines; instant iced tea (White Rose Redi-Tea); *Playboy* magazine; professional design magazine *Industrial Design* (today *I.D.*), New York.

Film: *Roman Holiday* with Hepburn and Peck, on a "Vespa" motor scooter in one scene, revealing Hollywood's enthusiastic adoption of the "Italian style"; Clouzot's *The Wages of Fear*; Zinneman's *From Here to Eternity*.

Literature: Baldwin's *Go tell it on the Mountain* and Miller's "The Crucible."

Theater: Beckett's "Waiting for Godot," exploring the search for God; Borodin, Wright, and Forrest's musical "Kismet."

French workers' strike against government's covert austerity measures crippled the country.

Workers rose up against the state in East Berlin which was subsequently overrun by Soviet tanks.

Raoul Dufy, Eugene O'Neill, and Joseph Stalin died.

The father of the ballpoint pen may have been John J. Loud of Massachusetts; he patented a complicated version in 1888. Equally impractical versions were subsequently patented in France, England, and America. Hungarian typographer and proofreader Lászlo József Biró (1889–1985), his chemist-brother Georg, and technician Imre Gellért incorporated the best features of past models and were granted a patent in 1938 in France and in 1945 in the U.S. But the outbreak of World War II thwarted production. Biró sold out to English banker H.G. Martin in 1944, who in turn sold rights to a Turin-born baron, Marcel Bich, and associates of Bich. Bich began making ballpoint parts in 1948, and produced the first French ballpoint pen in 1949. Seeking a non-disposable, high-quality version, the Parker Pen Company had developed the "Jotter" by 1953, according to Stuart L. Schneider (*The Incredible Ball Point Pen*, 1998). One hundred trial samples were given to employees, and then 30,000 demonstrator models were distributed to stores. The pen was an immediate success but the most expensive of any other at a price of $2.75. The "Jotter" included unprecedented features including a replaceable ink cartridge that lasted six times longer than a Bic; ink that became darker under light, dried fast, and did not smear or leak; and a button that, when pressed, rotated the cartridge 90°, saving wear on ball and socket. The "Jotter" acquired the reputation of being the best-engineered ballpoint pen ever made.

Date: 1954–56. Materials: Nylon, stainless steel, brass (example shown here of *c.* 1954). Manufacturer: Parker Pen Company, Janesville, WI, U.S. Photograph: © Stuart L. Schneider.

Events of 1954:

François Truffaut's essay "A Certain Tendency in French Cinema" legitimized film as an academic principle, arguing "authorship" belongs to the director, France.

Seagram Building was designed by Ludwig Mies van der Rohe, New York City (completed 1958).

Hochschule für Gestaltung building was designed by its director Max Bill, Ulm, Germany.

Premiers: The frozen TV dinner, by Swanson, U.S.; solar battery, by Bell Laboratories; nuclear-powered submarine (*Nautilus*); Compaso d'oro, an award for industrial-design excellence, Milan, Italy; "Univers" typeface was designed by Adrian Frutiger, Germany; Push Pin graphic design studio, organized by Milton Glaser, Seymour Chwast, and Edward Sorel, New York; successful organ transplant in a human, by surgeons at Peter Bent Brigham Hospital, Boston.

Films: Clouzot's *Diabolique*, Fellini's *La Strada*, *Rebel Without a Cause* with James Dean who subsequently crashed his Porsche and died (1955), Kurosawa's *The Seven Samurai*, Hitchcock's *Rear Window*.

Music: Britten's opera "The Turn of the Screw" and Menotti's opera "The Saint of Bleeker Street."

Literature: Golding's first novel *Lord of the Flies*, Murdoch's *Under the Net*, Sagan's *Bonjour Tristesse*.

Theater: Rattigan's "Separate Tables," Nash's "The Rainmaker," Adler and Ross's musical "The Pajama Game."

Dictator Abdal Nasser took full power of Egypt.

Pioneering photojournalist Frank Capra died from a land-mine accident, Vietnam. Colette, André Derain, Charles Ives, Auguste Lumière, and Henri Matisse also died.

The Finnish designer Timo Sarpaneva (b. 1926) was active in a number of areas including ceramics, textiles, and metalwork. After studying graphic arts in Helsinki, he became the head of the exhibition department and an artist at the Iittala glassworks in 1950. In the mid-1950s, he designed his first utilitarian domestic glass collection (represented by the examples here). For the series, he designed the symbol of a lowercase "i" in a circle which was subsequently adopted by Iittala as its trademark. The "i" series decanters were issued in a range of sizes from tall and narrow to short and round, with diminutive spouts and in a range of colors: green, light blue, lilac, and gray. The matching drinking glasses featured super-thin rims to make drinking more pleasurable. The absence of handles on the decanters was a common feature in Scandinavia at this time. Sarpaneva's sublime "i" forms showed more restraint than some of the more exuberant designs of the north countries that marked Scandinavian Modern as the quintessential expression of the 1950s. After establishing his own studio in 1962, Sarpaneva designed plastics for Ensto and glass for Corning in America and Venini in Italy. An innovator of techniques and forms, he worked in his clients' factories in order to gain mastery of processes and learn from their technicians.

Design: *c.* 1955. Material: Glass. Manufacturer: Iittala Lasitehdas, Iittala, Finland (from 1956–57, to 1960 for gray, to 1966 for all others). Photograph: Courtesy Montreal Museum of Decorative Arts (DB6.137.1), gift of Geoffrey N. Bradfield. Photo: Giles Rivest, Montreal.

Events of 1955:

"The Family of Man" exhibition was organized by Edward Steichen at The Museum of Modern Art, New York.

Premiers: The polio vaccine, by Dr. Jonas Salk; flight of *La Caravelle* airplane, France; Disneyland, near Los Angeles; Play-Doh children's toy, by Kenner; Scrabble board game, by Selchow & Richter; *The Guinness Book of World Records*; Mercedes 6-cylinder "Gullwing" auto; Thorazine (powerful tranquilizer) for mental patients; atomic clock, by Columbia University scientists; guided missile with an atomic-bomb warhead and jet fighter ("F-7-1") with air-to-air guided missiles, U.S.; patent of Velcro (inspired by plant burrs in 1948), by Georges de Mestral, Switzerland.

Film: Olivier's *Richard III*, Bergman's *Smiles of a Summer Night*, Hitchcock's *To Catch a Thief* with Grace Kelly.

Literature: Nabokov's *Lolita*, de Chardin's masterwork *The Phenomenon of Man*, Robbe-Grillet's *Le voyeur*, anthropologist Lévi-Strauss's *Tristes tropiques*, Kafka's *The Trial*, Wilson's *The Man in the Gray Flannel Suit* in sharp contrast to Ginsberg's poem "Howl," which signaled the emergence of the new "beat" generation.

Music: first recording of country-western singer Johnny Cash, Babbitt's "Two Sonnets," Adler and Ross's musical "Damn Yankees." Marian Anderson became the first African-American to sing at the Metropolitan Opera, New York.

"Inquiry Keyboard" was designed by John Schulte for the UNIVAC computer by Remington Rand, U.S.

Berlin-Moscow-Peking (Beijing)-Hanoi railway was inaugurated.

Edgar Faure became premier of France.

Albert Einstein, Alexander Fleming, Fernand Léger, Nicholas de Staël, and Maurice Utrillo died.

"Phonosuper SK 4" by Hans Gugelot and Dieter Rams

Dutch-Swiss architect and industrial designer Hans Gugelot (1920–65) studied architecture in Switzerland. From 1946–54, he worked with Max Bill and designed his first furniture, produced by Horgen-Glarus. As head of product design at the distinguished architecture-design school in Ulm, he espoused that a product should function efficiently without disguise or decoration. One of his pupils there, Dieter Rams (b. 1932), later became a collaborator in executing austere, functional forms for Braun, for whom Gugelot designed until his early death. Founded in 1921 in Frankfurt, Braun made industrial and scientific equipment at first and components for radios and record players later but, in 1929, began producing whole products. In 1936, Max Braun made his first battery-powered portable radio. When the sons of the founder, Artur and Erwin Braun, assumed management of the firm in 1951, Artur embarked on a new program that was to speak the new design language. Initially, Artur Braun and Fritz Eichler designed the products until Eichler developed a relationship with the students and staff of the Ulm school. In 1955, the entire radio line was redesigned in six months. The "Phonosuper SK 4" was an elegant expression of the new approach. Vents for sound emission made a highly considered, far-from-random geometrical pattern on the flat metal facade. The phonograph-radio unit was nicknamed *Schneewittchensarg* (Snow White's coffin), an amusing reference to the unadorned white box.

Date: 1956. Materials: Wood, metal, plexiglass. Manufacturer: Braun AG, Frankfurt am Main, Germany. Photograph: Courtesy The Museum of Modern Art, New York. Gift of the manufacturer. Photograph © 1999 The Museum of Modern Art.

Events of 1956:

Prince Rainier III of Monaco and former movie star Grace Kelly were married.

Construction began on the Guggenheim Museum by Frank Lloyd Wright, New York (to 1959).

Ionel Schein's all-plastic model house was shown at the Salon des Arts Ménagers, Paris.

Mary Quant opened a boutique, London.

Premiers: A successful video-tape recorder (VCR), by Charles Ginsburg at Ampex, U.S.; Fiat "600" and "Nuova 500" autos, Italy; transatlantic telephone cable; Pampers disposable diapers; famous lounge chair and ottoman in plywood and black leather, by Charles and Ray Eames for the Herman Miller Furniture Company, U.S.

Pop music made its début with the first television appearance of Elvis Presley, U.S.

British artist Richard Hamilton's collage *Just What Is It that Makes Today's Home so Different, so Appealing?* marked the artist as the father of Pop art.

Film: Ray's *Pather Panchali*, Morse and Honda's *Godzilla*, Bergman's The *Seventh Seal*, Preminger's *Man with the Golden Arm* with titles by Saul Bass.

Music: Bernstein's operetta "Candide," Lerner and Loewe's musical "My Fair Lady," Poulenc's opera "Les dialogues des Carmélites." Opera diva Maria Callas débuted in "Norma."

A milestone in British culture, Osborne's play "Look Back in Anger" was produced at London's Royal Court Theatre.

The ocean liner *Andrea Doria* was hit by the *Stockholm* and sank, resulting in legal claims of $40,000,000.

Conflict over the Suez Canal between Egypt and French-British forces resulted in its staying in Egypt's hands.

Clarence Birdseye, Bertolt Brecht, and Irène Jollot-Currie died.

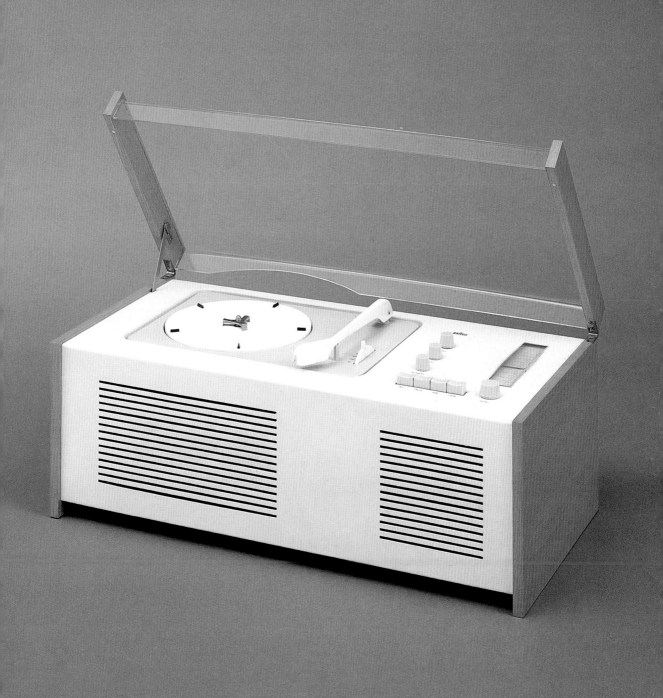

"Mirella" sewing machine by Marcello Nizzoli

Marcello Nizzoli (1887–1969) studied architecture and graphics in Parma, Italy, before moving to Milan. A painter at first, he became involved with the design of fabrics, exhibitions, and graphics. In a successful but brief association, from 1934–36, Edoardo Persico and Nizzoli designed remarkable exhibitions. Nizzoli was hired by Adriano Olivetti in 1936 as the first, and eventually the most successful, product designer of Olivetti office machines that have since become classics. His best industrial design for Olivetti appeared after 1945 when the firm began to reach customers worldwide. After the war, American financial aid was being used to help Italy ensure its position in the European marketplace. And, to achieve some presence, a number of Italian manufacturers in the decade 1955–65 collaborated with professional designers like Nizzoli. His abilities as a sculptor were obvious in his industrial design for clients such as Necchi. For the "Mirella" sewing machine he paid loving attention to all details including the precise placement of controls and the location of surface components. The equally innovative machinery that was hidden beneath Nizzoli's organic body was built with aluminum parts. The smooth, visually integrated housing was made possible by die-casting techniques that had been developed by Italy's thriving steel industry of the time.

Date: 1957. Materials: Enamel-painted steel, aluminum. Manufacturer: Necchi, Pavia, Italy. Photograph: Courtesy Die Neue Sammlung, Munich, Germany, inv. no. 38/68.

Events of 1957:

First artificial satellites *Sputnik I* and *Sputnik II* with the dog Laika were launched by the Soviets who also successfully sent a "super long distance" intercontinental ballistic rocket into space.

Architect Lucio Costa presented his plans for Brasilia, the new capital of Brazil (see p. 130).

Cadillac body by Harley Earl with exaggerated tail fins and rocket-like rear lights spoke of the quintessential, if dubiously tasteful, American dream car.

Cours Supérieur d'Esthétique Industrielle, the first school of design in France, opened, Paris.

Premiers: "Helvetica" typeface, by Edouard Hoffman and Max Miedinger, Switzerland; *Alouette II* helicopter, by Sud Aviation, France; Pontiac "Firebird III" prototype auto body, by Harley Earl, U.S.; Citroën "DS" auto, France; International Council of Societies of Industrial Design; Darvon, by Eli Lilly; frisbee sports disk (Pluto Platter); portable electric typewriter, by Smith Corona, U.S.; German nuclear reactor; underground nuclear testing, Nevada, U.S.; Wankel rotary motor, eliminating pumping pistons and rods, by Felix Wankel, Germany.

Film: Lean's *The Bridge on the River Kwai*, de Orduna's *El ultimo cuple*, Antonioni's *El Grido*.

Literature: Barthes's *Mythologies*, Shute's *On the Beach*, Robbe-Grillet's *Jealousy*.

Theater: Versatile Leonard Bernstein's musical "West Side Story," an update of "Romeo and Juliet," opened on Broadway; Jean Genet's gritty play "Le Balcon" premiered in London.

Pinyin was introduced to anglicize Chinese, but later changed.

Treaty of Rome established the European Economic Union.

Constantin Brancusi, Christian Dior, Jean Sibelius, and Auturo Toscanini died.

Trained as an architect under Kay Fisker, Arne Jacobsen (1902–71) was a complete designer. For the Royal Hotel of 1958–60 in Copenhagen, he designed practically every visual element, a tradition that had been established by others from the turn of the century including Wright (see p. 22) and Hoffmann (see p. 28). But *Gesamm-kunstwerk*, or a total-work-of-art, as it was earlier known, may never have been so immaculately conceived in the post World War II period as by Jacobsen. For the Royal Hotel, his most complete architectural piece, he designed the furniture, lamps, fabrics, glassware, eating utensils, dinnerware, and even door handles. Jacobsen's now-famous chairs, the "Egg" (shown here) and the "Swan," made their first appearance in the hotel. But the manufacture of the chairs would not have been possible had there not been the development of a new technology. In the mid-1950s, Danish furniture manufacturer Fritz Hansen acquired the rights to a new method of molding the inside of a chair as one continuous shell. With this knowledge, Jacobsen set about creating chairs that would exploit the method. He made prototypes in the form of plaster models, like sculpture, in his garage-turned-workshop. Calling on the traditional wing-chair model for his unique "Egg" concept, Jacobsen fully integrated the arms and body. The success of the "Swan" and "Egg" is beholden to the happy marriage of technology and aesthetics.

Date 1957–58. Materials: Fiberglass, foam, fabric, aluminum, PVC. Manufacturer: Fritz Hansen A/S, Allerød, Denmark. Photograph: Courtesy Fritz Hansen A/S.

Events of 1958:

Exposition universelle et internationale de Bruxelles featured the "Atomium" sculpture as an optimistic futuristic symbol, Brussels (41,000,000 visitors).

Wernher von Braun, former head of the German "V-2" rocket program and SS officer, became chiefly responsible for launching America's first satellite, *Explorer I.*

Roger Tallon designed the Gallic Tower (Tours Gallic) for La Coupe de Monde soccer finals, France.

Premiers: The stereophonic record; Sweet 'N' Low artificial sweetener; American Express Card; BankAmericard (later VISA); video game (ancestor of Pong), by William Higinbotham, U.S.; Alfa Romeo "Carabo" auto body, by Bertone, Italy; commercially available phototypesetting system (Diatype), by Berthold; commercially available color video tape; Pizza Hut restaurant, U.S.; integrated circuit, patented by Jack S. Kilby; NASA (National Aeronautics and Space Administration), Houston, Texas; Mr. Clean symbol for cleaning liquid (becoming known as Monsieur Propre in France, Don Limpio in Spain, and Maestro Lindo in Italy); voyage under the North Pole, by the nuclear-powered submarine *Nautilus.*

World's longest suspension bridge, the Mackinac, Michigan.

American jeweler Harry Winston donated the 45.5-carat Hope diamond, considered to hold a curse for the owner, to the Smithsonian Institution, U.S.

Film: *Mon Oncle* with Jacques Tati, Chabrol's *Le beau Serge*, Baker's *A Night to Remember*, Monicelli's *I Soliti Ignoti*, Eisenstein's *Ivan the Terrible II* (posthumous release).

Leo Castelli, a failed manufacturer turned art dealer, showed the art of Jasper Johns to phenomenal success, New York.

Literature: Pasternak's *Dr. Zhivago*, Pinter's *The Birthday Party*, Hasberg's *A Raisin in the Sun*, W.C. Williams's *Williams.*

Pope John XXIII replaced the late Pope Pius XII, Vatican.

In the second half of the 1950s, Hans Roericht (b. 1933) studied at the renowned architecture-design school in Ulm, Germany. He taught briefly in the U.S. as well as at his alma mater in Ulm, a town where he set up his own design studio in 1968. Roericht specialized in systems and environmental design. For Lufthansa he designed interiors, furnishings, and graphics and for the 1972 Munich Olympic Games stadium seating and desk systems. While at the Ulm school as a student, his diploma project had been a set of dinnerware. Inventing new dinnerware is a little like reinventing the wheel. And, besides, who needs a new one? At the Ulm school, Roericht's instructors—Hans Gugelot and others—had espoused the merits of systems design (see p. 136), and Roericht approached the dinnerware problem as if it were a system like any other, in which every element and adjunct function is a relative of the whole. The cup was a basic cylinder, the bottom narrower than the top, permitting stackability. In fact, all pieces were shaped for their stacking facility. Even though the pouring vessels, like pitchers, did not conform to the vertical geometry of other pieces, their forms offered relief within the system. The "TC 100" is still in production in the original white-glazed, high-density, heavy-weight porcelain.

Date: 1959. Material: White-glazed porcelain. Manufacturer: Thomas Porzellan-Werk (today, part of Porzellanfabrik Rosenthal), Waldershof/Oberpfalz, Germany. Photograph: © Greg S. Krum, New York.

Events of 1959:

Women's suffrage was barred by voters, Switzerland.

The Soviet Union's spacecraft *Luna 3* transmitted the first photographs of the backside of the Moon.

Finnish architect Alvar Aalto designed a satellite community according to principles of a "bedroom community" whose transportation access was the primary concern, Bremen-Vahr, Germany.

Premiers: Panty hose; Barbie doll ($3), by Mattel, U.S.; commercially available photocopying machine, by Xerox, U.S.; Austin's "Morris Mini 1000" auto body, by Alex Issigonis, U.K.; Fiat "500" auto body, by Dante Giacosa, Italy; crash testing of autos, by Bella Barenyi at Mercedes, Germany; "1401" computer, by IBM.

Joan Miró painted murals for the UNESCO building, Paris.

Escuela Massana (school of arts and crafts) began courses in graphic design, Barcelona.

Film: Goddard's *A bout de souffle*, introducing the nouvelle-vague technique into film making; Mario Camus's *Black Orpheus*; Chukhral's *Ballad of a Soldier*; Richardson's *Look Back in Anger*; Wilder's *Some Like It Hot* with Monroe.

Literature: Burroughs's *Naked Lunch*, Ionesco's "Rhinoceros," Fleming's *Goldfinger*, Grass's *The Tin Drum*.

Music: Rogers's musical "The Sound of Music," Thomson's "Collected Poems," popularity of Brecht's "Mack the Knife" (1928). Joan Sutherland triumphed in the opera "Lucia di Lammermoor."

Threatened by China, the Dalai Lama fled into exile in India.

Fidel Castro and *los barbudos* (the bearded ones) took control of Cuba; dictator Bastista had fled the day before with most of the country's treasury.

Bernard Berenson, Cecil B. DeMille, Jacob Epstein, and Frank Lloyd Wright died.

Stacking chair by Verner Panton

Verner Panton (1926–1998) studied architecture in Odense, Denmark, in the mid-1940s and, until 1951, at the Royal Academy of Fine Arts in Copenhagen. He traveled widely and eventually opened an architecture-design office in Copenhagen. In the late 1950s, Panton joined the 20th-century-long quest for a one-piece chair that would require no assembly (see p. 142). He began experimenting with the kind of plastics technology initiated in the airplane industry. In 1960, he succeeded in making a chair in polyurethane foam, but it was not mass produced due to technological difficulties, the current lack of interest in plastic furniture, and the high cost of dies and machinery. The undaunted Panton and Manfred Diebold, a Vitra-furniture-company plastics expert, had by 1963 developed a method for making fiberglass-polyester prototypes, based on Panton's plaster models. It marked the advent of the first one-piece continuously formed chair able to be produced on a large scale. When manufacture began in 1967, the material was changed to deformed-and-lacquered rigid polyurethane foam (Baydue) and in 1970 was further changed to a non-fiberglass plastic (Luran-S). Today, the chair is produced in a structural, integral foam (PUR). Its smooth, lustrous finish and flowing lines are reminiscent of an automobile fender.

Date: 1960. Material: Luran-S thermoplastic (example shown here). Manufacturer: Vitra GmbH, Basel, Switzerland (currently Weil am Rhein, Germany) (from 1967), for Herman Miller International. Photograph: Courtesy Montreal Museum of Decorative Arts, gift of Geoffrey N. Bradfield (D84.171.1). Photo: Giles Rivest, Montreal.

Events of 1960:

Brasilia, the new capital of Brazil, praised by architect Le Corbusier for its "spirit of invention" and by André Malraux as "The Capital of Hope," was built to the 1956 plans of Lucio Costa, Oscar Niemeyer, and others.

Convent of La Tourette was built by Le Corbusier, France.

The Soviets felled U.S. pilot Francis Gary Powers's reconnaissance flight of the Lockheed "U-2" spy airplane. (Powers was exchanged in 1962 for Soviet spy Rudolph Abel.)

Premiers: The all-transistor TV, by Sony; U.S. approval of the oral polio vaccine by Albert Sabin; heart pacemaker, U.K.; Librium, by Roche Labs; weather satellite (*Tiros I*); nuclear-powered aircraft carrier (*Enterprise*); Letraset letter-transfer typography; "U2" spy airplane, by Clarence L. Johnson, U.S.; launching of ocean liner *France*; effective oral birth-control pill, by endocrinologist Gregory Goodwin Pincus, U.S.; operational laser, by Theodore Maiman at Hughes Research Labs, Miami; OPEC (Organization of Petroleum Exporting Countries).

Film: Antonioni's *L'Avventura*, Kubrick's *Spartacus*, Preminger's *Exodus*, Hitchcock's *Psycho*.

Literature: Shirer's *The Rise and Fall of the Third Reich*, Camus's *The Plague*, Bolt's *A Man for All Seasons*, Banham's *Theory and Design in the First Machine Age*.

Theater: Bart's musical "Oliver!," Schmidt and Jones's musical "The Fantasticks," Hellman's "Toys in the Attic."

Joseph Kittenger Jr. made a 90,000-foot parachute jump, remaining the highest since, in 4.5 minutes.

Sharpeville, South Africa, massacre by the national police became a symbol of apartheid's brutality (69 killed).

Adolf Eichmann, overseer of the Holocaust during World War II, was captured by Israeli secret agents, Buenos Aires.

Albert Camus, Clark Gable, and Boris Pasternak died.

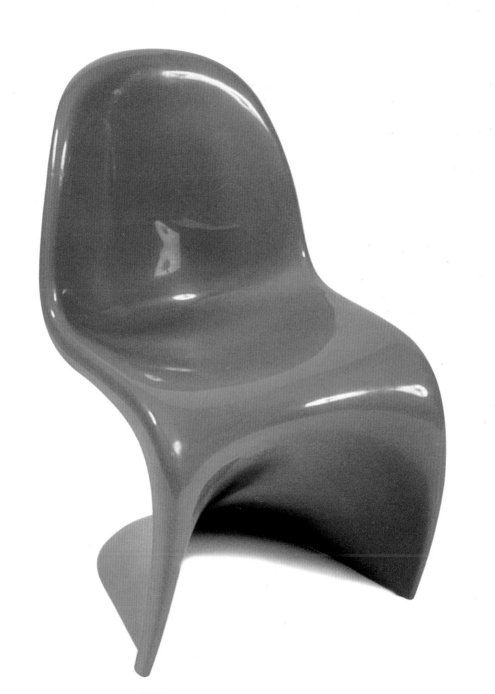

Encouraged by the exciting theories espoused by architect Le Corbusier in his book *Vers une architecture* of 1925, Eliot Noyes (1910–77) studied under and then worked for former Bauhaus head and teacher, Walter Gropius and Marcel Breuer, in Cambridge, Massachusetts, in the late 1930s. Fortune came his way in the early 1940s when he was appointed the first director of industrial design of New York's Museum of Modern Art. While there, he organized some seminal, high-profile exhibitions. After the war, in Norman Bel Geddes's office, Noyes designed an electric typewriter for IBM. When the Bel Geddes office closed, IBM head Thomas Watson Jr. retained Noyes to manage a corporate-design department based on the approach of the Italian firm Olivetti. In the late 1950s, IBM engineers developed a revolutionary typing mechanism for which, of course, Noyes designed the case. Known as the "Golf Ball 72" and later officially called the "Selectric," the writing machine was the first single unit to offer interchangeable typefaces (in the form of snap-in balls) and a carbon-plastic ribbon. It weighed 31 lbs. (14 kg). The machine's sculptural lines were unabashedly influenced by the work of the Olivetti designer Marcello Nizzoli (see p. 126). The bright color range of the "Selectric" machine was unusual for office equipment of the time, particularly in America.

Date: 1961. Materials: Painted and polished steel, plastic. Manufacturer: IBM, U.S. Photograph: IBM Archives, Somers, NY, U.S.

Events of 1961:

Apelmann symbols of a man for "stop" and "go" traffic signals were designed by Karl Peglau, East Germany.

Architect Le Corbusier built the Maison de la Culture, Firminy, France.

Premiers: The electric toothbrush, U.S.; Salone del Mobile (furniture fair), Milan, Italy; a wall erected between East and West Berlin thus creating Eastern and Western Europe; artificial coffee lightener (Coffee-Mate); Weight Watchers program, U.S.; Valium pharmaceutical tranquilizer; Amnesty International and architectural group Archigram, both London; CNES (national center for space studies), France; Elisava design school, Spain.

Film: Truffaut's *Jules and Jim*, Visconti's *The Damned*, Renais's *Last Year at Marienbad*, Wise's *West Side Story*.

Literature: Miller's *Tropic of Cancer*, Heller's epic satire *Catch-22*. Spark's *The Prime of Miss Jean Brodie*, Percy's *The Moviegoer*, Salinger's *Franny and Zooey*, Naipaul's *A House for Mr. Biswas*.

Music: Harris's "Canticle of the Sun," Moore's "The Wings of the Dove," Mancini and Mercer's song "Moon River."

The U.S. launched the space capsule *Mercury* with the chimpanzee Ham. Soviet cosmonaut Yuri Gagarin became the first human sent into outer space. President Kennedy committed the U.S. to putting a person on the Moon within a decade.

Russia exploded its first hydrogen bomb, far larger than the first by the U.S. in 1952.

South Africa left the British Commonwealth to become independent.

Luigi Einaudi, Ernest Hemingway, Chico Marx, Eero Saarinen, and Max Weber died.

In 1931 in America, Joseph Schick introduced the first man's electric shaver. By 1936, sales of all electric shavers had reached the first million unit. A Schick shaver designed by Raymond Loewy (see p. 80) was selling for the hefty price of $15 in 1941. The Braun company (see p. 124) developed a model in 1938 with a simple case and a battery-rechargeable oscillating motor inside. Unable to be produced until after the war, it was introduced in 1950 as the model "S 50." Regardless of its advanced technology, the design had become dated. By 1954, Braun had caught up aesthetically and made an arrangement to share its technology with the Ronson Company in America, thus assuring its investments on both sides of the Atlantic. But Ronson's level of design was far inferior to that of Braun's. When Max Braun died in 1961, a new regime, headed by his sons, transformed the firm's product-design program and corporate-identity graphics (see p. 124). From 1954–58, Hans Gugelot (1920–65), one of its principle designers, brought demurity to Braun products, made only in black or white materials. For the "Sixtant" shaver, a new process was used to produce the foil head. Gugelot's thinking was stringently founded on the principle of the system in which every particle is part of the greater whole, and all aspects exist in tandem with one another. Many of his products, like the razor here designed with Dieter Rams (b. 1932) and Gerd Alfred Müller (b. 1932), have retained their youthful good looks after a quarter century.

Date: 1962. Materials: Plastic, steel. Manufacturer: Max Braun, Frankfurt, Germany. Photograph: Courtesy Die Neue Sammlung, Munich, Germany, inv. no. 13/68. Photo: Sophie-Renate Gnamm.

Events of 1962:

The Americans and Soviets dangerously confronted one another, with the threat of a nuclear war, over ballistic missiles being placed on the island of Cuba by the Soviets.

The New National Museum was designed by Ludwig Mies van der Rohe, Berlin (completed 1967).

Yves Saint-Laurent fashion house was opened, Paris.

Premiers: The industrial robot, by Unimation Inc., U.S.; Lear jet; folding bicycle, by Alex Moulton, U.K.; a version of the Kodak "Carousel" slide projector, designed by Hans Gugelot, Germany; Lufthansa airlines graphics program, still implemented today, by Hans Roericht and Otl Aicher, Germany; Kmart and Wal-Mart stores, U.S.; launch of telecommunications satellite (*Telstar*), by AT&T.

Film: Welles's *The Trial*, *Cleopatra* with Elizabeth Taylor and Richard Burton, Lean's *Lawrence of Arabia*.

Literature: Lévi-Strauss's *La pensée sauvage*, Burgess's *A Clockwork Orange*, Baldwin's *Another Country*, Albee's play "Who's Afraid of Virginia Woolf?", Solzhenitsyn's *One Day in the Life of Ivan Denisovich*, McLuhan's *The Gutenberg Galaxie*, Marighella's handbook *Mini-Manual of the Urban Guerrilla*, instructions for causing mass unrest.

Music: Beatles's first song "Love Me Do" and Britten's "War Requiem."

Thalidomide, advertised as the "sleeping pill of the century," deformed the babies of pregnant women; however, it later proved effective in treating leprosy.

John H. Glenn became the first American to orbit the Earth.

The U.S. and the Soviets came to the brink of war over missiles in Cuba.

e.e. cummings, Yves Klein, Franz Kline, Charles Laughton, and Marilyn Monroe died.

Olaf Bäckström (b. 1922) worked as an engineer during the first decade of his career. In 1954, he turned to woodcarving. Adept but without formal training, he made wooden household objects whose excellence was recognized by a silver medal at the Milan Triennial of 1957. From 1958, Bäckström worked as an industrial designer at Fiskars in Helsinki. His initial assignment there was a tableware collection in plastic. His flatware for camping received another silver medal at the Triennial in 1960. Far better known than its designer who lacked the education expected from such a sound thinker, Bäckström's orange-colored handles on the "O-Series" scissors and shears have become recognizable the world over. He began the project by carving prototypes in wood, which is understandable considering his early dedication to the medium. Ergonomic brass-handled scissors had been used by tailors for years, if not centuries. Drawing on these models, prototypes were made in 1960–61 with blue molded handles (shown here). The blades, which were too expensive to make, were reconceived using a new technology and later were changed once again to further save costs. The final model was introduced in 1967. The blue handles were changed to orange, a fashionable color in the late 1960s. Unfortunately, success has bred shameless copies by others. With a distinguished body of work behind him, Bäckström left Fiskars in 1980.

Date: 1960–63. Materials: ABS resin, steel (example of 1960–61 shown here). Manufacturer: Fiskars Oy Ab, Billnäs, Finland. Photograph: Courtesy Fiskars.

Events of 1963:

The first woman (Valentina V. Tereshkova) in space was launched by the Soviets.

Mini-skirts were introduced by fashion designer Mary Quant, Britain.

The first injection-molded chair ("Polyprop"), by Robin Day in 1954, was produced by Hille & Co. in great quantities, U.K.

Artist César presented a transparent TV set at "L'Objet" exhibition at the Musée des Arts Décoratifs, Paris.

Premiers: "Air Force One" (the U.S. president's airplane), exterior motif and interior by Raymond Loewy; an auto produced in Brazil, designed by Brooks Stevens; tear-strip opener for soda-pop can, patented by Ermal Fraze of Reliable Tool & Machine Co., U.S.; practical holography, at the University of Michigan, U.S.; minicassette recording tape, by Philips, Netherlands; practical applications of foam rubber and molded polyester; process for producing soft eye lenses, by Otto Wichterle, Czechoslovakia (patent bought by U.S. optometrist Robert J. Morrison and ultimately by Bausch & Lomb); prenatal blood transfusion, New Zealand; ZIP mailing codes, U.S.; "Instamatic" camera (with film cartridge), by Eastman Kodak; lung transplant in a human; touch-tone telephone.

Film: Richardson's *Tom Jones*, Visconti's *The Leopard*, Regueiro's *El buen amor*, Shepitko's *Heat*.

Literature: McCarthy's *The Group*, Böll's *The Clown*, Sinyaski's *Fantastic Stories*, Clézio's *The Interrogation*.

Lejaren Hiller composed the "Computer Cantata."

Design courses began at the ENSAD design school, Paris.

U.S. President John F. Kennedy was assassinated. Georges Braque, Jean Cocteau, Edith Piaf, Francis Poulenc, and Tristan Tzara also died.

In the late 1930s, Marco Zanuso (b. 1916) studied architecture in Milano at the Politecnico where he became a teacher and eventually the director of its architectural faculty. He set up a practice at the end of the war and served as the editor of the journals *Domus* and *Casabella*, eminent commentators on the state of architecture and design. Zanuso's "Lady" chair of 1951, a demonstration of a pioneering spirit, was one of the first to replace traditional seating construction with foam. The Pirelli-developed material encouraged the streamline forms of the day. Zanuso's technological approach to rational solutions, different from that of some others in Italy, was realized in his work for Kartell (see p. 142) and Brionvega. The most courageous of the firms which made audio-visual equipment, Brionvega executed a number of products by Zanuso, who collaborated with Richard Sapper from 1958–77. Their "black box" TV set (the "Black 201") of 1969, the "Ts 502" radio, and equipment for others achieved a cult status in Italy in the 1970s. The little "Ts 502" became a secret object when closed; opened, it exposed a speaker in one side and technical-looking controls in the other. The handle and automobile-type antenna were extendible. Since the heavy electronics were located in only one of the halves of the cube, the handle was precisely placed off-center for balance during transit.

Date: 1964. Materials: Painted die-cast alloy (case), zama plastic (inside fascia). Manufacturer: Brionvega, Milan, Italy (1964–1978). Photograph: © Gregory S. Krum, New York. Collection of Form and Function, New York.

Events of 1964:

President Johnson was able to escalate America's attack on North Vietnam without the legal and political complications which a formal declaration of war would have entailed.

The first Habitat store was opened by Terence Conran, London.

Based on a joke he suggested, American designer Rudi Gernreich introduced the monokini, a topless bathing suit for women, U.S.

A gasoline pump for Mobil Oil was designed by Eliot Noyes.

Premiers: Ferrari "275 GTB4" Berlinetta, Italy; Ford "Mustang" auto; G.I. Joe doll for boys; Bullet Train (Tokyo-Osaka, Japan); integrated-circuit computer, by IBM, U.S.; controllable canopy (Ram air canopy) parachute, by Jean Gilbert, France; "BIO" (Biennial of Industrial Design/Bienale Industrijskega Oblikovanja), Ljubljana, Yugoslavia.

Film: Cukor's *My Fair Lady* and Stevenson's *Mary Poppins*.

Literature: Achebe's *Arrow of God*, Dahl's *Charlie and the Chocolate Factory*, McLuhan's *Understanding Media*, Marcuse's *One-Dimensional Man*.

Music: Bock and Harnick's musical "Fiddler on the Roof"; musical "Funny Girl," comic Fanny Brice's biography, with Barbra Streisand; Stockhausen's "Mikrophonie I," scored for gong, microphones, and filters. The Rolling Stones released their first album, U.K.

Health-warning statements were added to cigarette packs.

Nikita Khrushchev was ousted as premier of the U.S.S.R.

Murray Gell-Mann and George Zweig proposed that protons and neutrons are made of smaller particles called quarks, and Oscar Greenberg and Yoichiro Nambu proposed that quarks come in different varieties called colors (to 1965).

Brendan Behan, Ian Flemming, Peter Lorre, Jawaharial Nehru, and Cole Porter died.

"4867" stacking chair by Cesare "Joe" Colombo

Cesare "Joe" Colombo (1930–71) studied art in the late 1940s and subsequently attended the Politecnico in Milan. During the early 1950s, he was principally an avant-garde painter and sculptor of the arte povera school. He turned his interests to design and architecture in the mid-1950s, took on his family's electrical-equipment business at the end of the decade, and set up a design office in 1962. Even though he was active in design for barely a decade, dying at age 41, he was prolific, and his clients for furniture, pottery, lighting, and electrical appliances included Candy, Bernini, Stilnovo, and O Luce. Colombo became one of the stable of designers that Kartell founder Giulio Castelli was commissioning in the 1950s and '60s to design functional objects made of newly patented plastics. Castelli's efforts were helping to make plastics respectable for home use. Colombo's "4867" stacking chair in ABS resin for Kartell was the result of experiments with new technology and the century-old search for a one-piece chair. Counter to the concurrent work of others in Italy and elsewhere (see p. 132), the Colombo chair that had been intended for manu-facture in aluminum was supposed to be the first of its kind—a chair molded entirely of one mater-ial. Because there were difficulties with the molds and the process, full production was delayed until 1967. The legs had to be made separately, and the hole in the back of the seat was necessary for removal from the mold. Colombo's original name for the "4867" was "Universale."

Date: 1965. Materials: ABS resin (from late 1967), polypropy-lene (from 1976). Manufacturer: Kartell S.p.A., Noviglio (MI), Italy, and Easley, SC, U.S. Photograph: Courtesy Kartell S.p.A.

Events of 1965:

Aleksei Leonov left the Soviet *Voskhod 2* spacecraft that was orbiting the Earth and became the first human to "walk," but tethered, in space, for ten minutes.

"Action Office" furniture, an open-office-plan system by Ron Beckman, Leif Blodee, George Nelson, Robert Probst, and others, was produced by the Herman Miller Furniture Company, U.S.

Premiers: The domestic answering machine, by Robosonics, U.S.; IBM logo, by Paul Rand; Celanese logo, by Saul Bass; Unimark design studio, Milan; Volkswagen "Karmann-Ghia" auto body, by Carrozeria Ghia of Turin; Industrial Designers Society of America (the merger of Industrial Designers Institute and the American Society of Industrial Designers), U.S.; indoor sports stadium (Astrodome), Houston, Texas; totally new America mass-transit system (BART) since 1907, by Sundberg-Ferar, San Francisco.

Student agitation against the Generalissimo Francisco Franco's repressive regime was escalated, Spain.

Industrial-design department was formed at the Escuela Massana, Spain.

Donald Judd, Ad Reinhardt, and others emerged as pragmatic sculptors and painters of Minimalism.

Film: Olivier's *Othello*, Lean's *Dr. Zhivago*, *Help!* with the Beatles, *The Sound of Music*.

"Spirit of America" became the fastest auto to date to achieve an officially recorded speed of 995 km/h., U.S.

Tibetan Autonomous Region was established as part of China, essentially formally dissolving the Tibetan state.

T.S. Eliot, Dorothea Lange, Le Corbusier, Somerset Maugham, Albert Schweitzer, and Edgard Varèse died.

Trained in Italy and America, Gino Valle (b. 1923) worked in his father's architecture studio in Udine, Italy, before he and his brother took over the office. Like many architects in Italy, he became active in both architecture and industrial design. His clients included clock maker Ritz-Italora and, from 1954, appliance manufacturer Zanussi. On the heels of the development of a new analog system, Solari, located in Valle's native Udine, commissioned him to design a line of clocks and automated schedule boards. Giant versions of the boards began appearing in public spaces such as airline and railroad terminals, popular at the time but since replaced by video monitors. The electromechanical system utilized modular flaps. Numbers or whole words or phrases were split horizontally across the middle. Regulated by instructions from a central source (human or automatic), the top flaps of numbers or lettering, rotating along a wheel axis, dropped down. While the technology is interesting, the success of Solari's clocks lay in Valle's aesthetic contribution. The form of this version of the "Cifra 5" was determined by the circular nature of the mechanism. Clear and colored plastics, a finger-controlled adjustment wheel, and a stabilizing foot all retained the severe geometry of a cylinder. Graphic designer Massimo Vignelli (b. 1931) chose the typeface Helvetica for the easy-to-read, white-on-black numbers.

Date: 1966. Materials: ABS resin, methyl methacrylate, metal. Manufacturer: R. e C. Solari Udine S.p.A., Udine, Italy. Photograph: © Gregory S. Krum, New York.

Events of 1966:

Buildings, thousands of treasures, and 130,000 photo negatives were ravaged by devastating waters of the overflowing Arno River, Florence.

The Soviet *Luna IX* made the first soft landing on the Moon. U.S. troops were sent to Vietnam.

Radical-design movement began in Italy with the establishment of the Archizoom Associati (to 1974) in Florence, Superstudio in Milan, and Gruppo Strum in Turin.

"Les années 25" exhibition, from which "Art Déco" was first derived, was held at Musée des Arts Décoratifs, Paris.

Premiers: The color avocado in major electrical appliances, by General Electric; color video recorder ("Betamax"), by Sony, Japan; "Eurodomus" design fair, Genoa, Italy; MasterCharge (later MasterCard); fiber-optic telephone cables; "Star Trek" television series; Unimark design studio, organized by Massimo Vignelli and others, Milan, Italy.

Film: Pasolini's *Gospel According to St. Matthew*, Lelouch's *A Man and a Woman*, Bergman's *Persona*.

Art: A series of electric-chair images was made by Andy Warhol. Robert Indiana painted *Love*.

Music: Tippett's opera "The Knot Garden" and the musical "Cabaret" directed by Hal Prince. Barber's "Antony and Cleopatra" opened at the new Metropolitan Opera House building, New York.

Literature: Lin Biao's *Quotations from Chairman Mao* ("Little Red Book") in English, Masters and Johnson's *Human Sexual Response*, Greene's *The Comedians*, Sontag's *Against Interpretation*, architect Venturi's *Complexity and Contradiction in Modern Architecture* that led to the Postmodern tendencies of the next decade.

Storm of Leonid meteors, 20,000 per hour, swept over the Earth in the wake of the 55P/Temple-Tuttle comet.

Jean Arp, Vincent Auriol, Walt Disney, Alberto Giacometti, and Evelyn Waugh died.

"Blow" armchair by De Pas, D'Urbino, Lomazzi, and Scolari

The restlessness of the late 1960s, a time of throw-away aesthetics and the rejection of established norms, infected designers of the time. In 1967, a propitious proclamation by a group of young architects—Gionatan De Pas (1932–91), Donato D'Urbino (b. 1935), and Paolo Lomazzi (b. 1936)—argued that a fresh design voice closer to that of the new Pop culture was needed. Even though their early furniture ideas and prototypes were conjured with a blind eye to mass production, they explored low-cost furniture design that might have the kind of transformable, nomadic characteristics which pneumatic technology would help to realize. At a time when architectural inflatables were being built by others, De Pas, D'Urbino, and Lomazzi became the first to develop a practical pneumatic piece of furniture—the "Blow" chair. It was also the first inflatable Italian design object absent of a high price and élitist credentials. The solution was a distinct departure from the weight and substance of traditional upholstered furniture. The production of the "Blow" was made possible by a method of electronic welding that employed a radio frequency to seal the seams of the polyvinylchloride balloon. The chair, the first by the designers who have since become prolific, established both them and the manufacturer as innovators.

Date: 1967. Material: Clear or tinted calendered polyvinyl-chloride. Manufacturer: Zanotta, Nova Milanese (MI), Italy. Photograph: Courtesy Zanotta. Photo: Marino Ramazzotti.

Events of 1967:

Exposition universelle ("Terres des hommes/Man and His World") opened, Montréal (50,000,000 visitors).

Fire erupted on U.S. *Apollo 1* space capsule, killing astronauts Virgil I. Grissom, Edward H. White, and Roger Chaffee. *Saturn* deep-space rocket was realized through the efforts of Wernher von Braun and others at NASA.

Gruppo 9999 radical-design consortium was founded, Florence.

Premiers: The quartz watch; commercially available domestic microwave oven, by Amana, U.S.; French nuclear submarine (*La Redoutable*); successful human heart transplant, by Dr. Christian Barnard, South Africa; French color television broadcast; MIRV (multiple independent re-entry) space vehicle; PBS (Public Broadcasting System), U.S.

With outbreak of war in Nigeria and the secession of Biafra as a republic in 1967, more than 1,000,000 people had died by 1970.

Poet Alan Ginsberg organized the first "be-in" to harness the "flower power" and "good vibes" of hippies, San Francisco.

Film: Buñuel's *Belle du jour*, Pasolini's *Oedipus Rex*, Donen's *Two for the Road*, Nichols's *The Graduate*.

Music: The Beatles's "Magical Music Tour," Morrison's song "Light My Fire," Sessions's "Symphony No. 7."

Literature: Márquez's *One Hundred Years of Solitude*, Potok's *The Chosen*, Stoppard's play "Rosencrantz and Guildenstern Are Dead," McLuhan's *The Medium Is the Message*, expanding on "hot" and "cool" ideas in print and movies.

In the Six-Day War, Israel captured the Sinai Peninsula, Jerusalem Old City, West Bank, and Golan Heights.

Becoming the fifth nation with nuclear capability, China detonated an hydrogen bomb.

John Coltrane, "Che" Guevara, and René Magritte died.

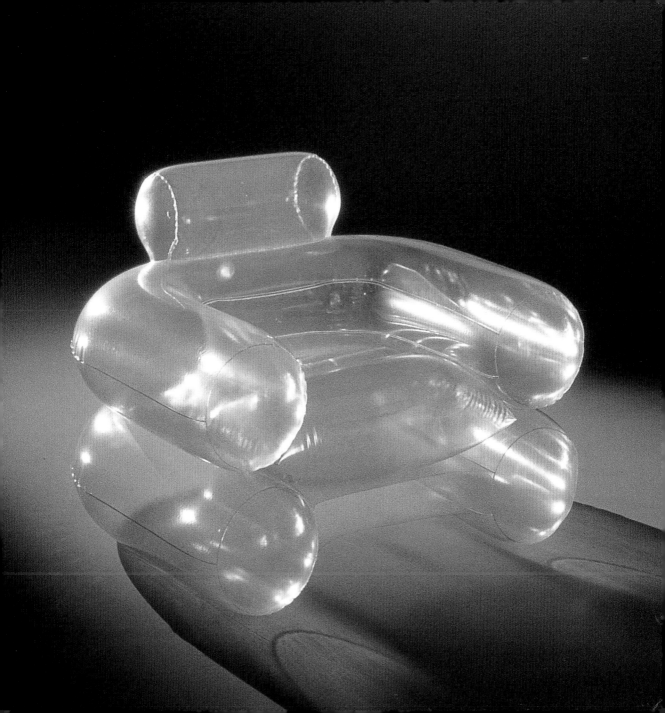

Olivier Mourgue (b. 1939) studied in Sweden, Finland, and his native France until 1961 when he established his own studio in Paris. His rich body of work included furniture, textiles, environments, and toys. Mourgue came to international prominence with the bright-red undulating furniture that was installed in the stark-white, and uninviting hotel lobby of the Stanley Kubrick film *2001: A Space Odyssey* of 1968. Mourgue's persisting interests have centered on flexible environments, multiple use, stretch fabrics, foam rubber, and organic forms. All were called into play for the chaise longue "Bouloum," Mourgue's nickname as a child. The first version of the shape—based on the body shape of one of Mourgue's friends—was formed by a metal frame; it was later replaced by a stiff molded-plastic slab. The portable, light-weight, and stackable seat could be randomly arranged indoors or, without the upholstery, outside. In 1970, Mourgue continued the "Bouloum" theme in his design of the French pavilion at the world's fair in Osaka, Japan. The figure was used for seating in restaurants and meeting halls and, in an upright position, as if standing to attention, for stanchions that were applied with informational and directional graphics. Mourgue's work of the 1960s helped to make French design accessible to the public by humanizing its lofty attitude of the time.

Date: 1968. Materials: Nylon jersey, polyurethane foam, polyester fiber. Manufacturer: Airborne, Toumus, France. Photograph: Courtesy Philadelphia Museum of Art, Philadelphia, PA.

Events of 1968:

Finnish-born American architect Eero Saarinen's Gate Way Arch rose gracefully above the Mississippi River, St. Louis.

U.S. astronauts of the space capsule *Apollo 8* became the first humans to orbit the Moon.

"Minimal style" was developed by Bang & Olufsen for its sound and image equipment, under the direction of designer Jakob Jensen, Denmark.

Premiers: The waterbed; artificial larynx; ocean liner *Queen Elizabeth II*, by Cunard; U.S. motion-picture decency ratings; album to sell "platinum" (1,000,000 copies of Cream's "Wheels of Fire"); Spanish singer Placido Domingo at the New York Metropolitan Opera.

The widow of U.S. president John F. Kennedy and Aristotle Onassis married.

Music: Beatles's "Hey Jude"; first rock musical "Hair"; invention of a space-age instrument, the Moog synthesizer, by Robert Moog.

Films: Kubrick's *2001: A Space Odyssey* by Arthur C. Clarke and *Lion in Winter* with Katherine Hepburn and Peter O'Toole.

Literature: Nehls's *The Sacred Cows of Functionalism Must Be Sacrificed*, Vidal's *Myra Breckenridge*, Oe's *A Personal Matter*, Wolfe's *Electric Kool-Aid Acid Test*.

Rioting students, led by Daniel Cohn-Bendit, froze transportation, the mail, and almost the telephone system, Paris.

French skier Jean-Claude Killy received gold medals in all Alpine events at the Winter Olympics, Grenoble, France. Army lieutenant Arthur Ashe became the first black person to win the U.S. Tennis Open.

The brief foray toward progressive reform, known as the Prague Spring, was oppressed by the Soviets, Czechoslovakia.

Martin Luther King Jr. and Robert F. Kennedy were assassinated. Marcel Duchamp and Yuri Gagarin also died.

In the late 1960s, student unrest and rebellion, rampant in the U.S. and Europe, was particularly violently expressed against imbedded, decadent political systems and policies. The youth-based anti-orthodoxy was being mirrored in furniture and furnishings in Italy where, in 1965, the team of Piero Gatti (b. 1940), Cesare Paolini (b. 1937), and Franco Teodoro (b. 1939) set up a design office in Turin. They began exploring how they might be able to create an "anthropological" object—one that would satisfy not only traditional requirements for comfort but also the fickle sensual interests of the restless youth culture of the time. Purportedly, in the designers' studio, a bag that was being used for dumping plastic garbage gave birth to the seating concept shown here. The realization was a low-cost, universal object that would conform "to any body, in any position, on any surface," in the designers' words. The stuffing, somewhat like the plastic refuse in the designers' office, consisted of 12,000,000 small plastic pellets, according to the manufacturer. No matter the radical nature of the concept, what forward-thinking manufacturer, especially an Italian one in the late 1960s, would not have produced a chair, even if it did not sell well, that demanded no new technology, expensive machinery, tooling, or dies? The "Sacco" indeed became a fashionable icon, a symbol of the epoch, and a phenomenal success.

Date: 1968–69. Materials: Leather, fabric, or Telafitta (poly-vinylchloride-coated fabric); expanded polystyrene beads. Manufacturer: Zanotta, Nova Milanese (MI), Italy. Photograph: Courtesy Zanotta. Photo: Foto Masera.

Events of 1969:

Stepping from the landing craft *Eagle* of the capsule *Apollo 11*, U.S. astronauts Neil A. Armstrong and Buzz Aldrin became the first humans to walk on the Moon.

Ettore Sottsass's ruby-red "Valentine" portable typewriter with its integral case was produced by Olivetti, Italy.

"Central Living Block" by Italian designer Cesare "Joe" Colombo for the "Visiona" exhibit at the furniture fair in Cologne was inspired by science fiction and featured plastic furniture in an open-living plan, Germany.

Premiers: Arpenet (pre-Internet), for linking military researchers' computers, by Jonathan B. Postel and others, for the U.S. Defense Department, Washington; jumbo jet (Boeing "747"); non-stop solo circumnavigation of the Earth by boat; Nobel Prize for economics; female graduates at Yale University, U.S.; frogdesign studio, by Hartmut Esslinger, Germany; Centre de Création Industrielle at the Centre Pompidou, Paris; "Sesame Street" children's TV program, U.S. (barred by the BBC in 1971); "Monty Python's Flying Circus" TV series, U.K.; flight of the *Concorde* supersonic airplane; quirky view of Post-modernism (SITE Projects buildings); food processor for commercial use, by chef Pierre Verdun, France.

Films: *Isadora* with Vanessa Redgrave, *Sundance Kid* with Redford and Newman, Russell's *Women in Love*, Fellini's *Satyricon*.

Literature: Nabokov's *Ada*, Fowles's *The French Lieutenant's Woman*, Puzo's *The Godfather*.

Chanel biographical musical "Coco" opened on Broadway, New York.

Woodstock Music Festival was first held, Woodstock, NY.

Belgian cyclist Eddy Merckx won the first of five victories he would achieve in the 2,400-mile Tour de France race.

Camille, at 210 m./h., became the strongest hurricane to reach land in the U.S. (southeastern coast).

Judy Garland, Walter Gropius, and Mies van der Rohe died.

Shiro Kuramata (1934–91) was trained as an architect and subsequently a cabinet maker. His shop interiors for the clothing designer Issey Miyake were better known than Kuramata himself. In fact much of his furniture may be élitist in that it is expensive and little known by the public. For example, the cabinet here costs about $10,000. Yet, his work became tremendously influential in professional design circles. The chest shown here, like much of Kuramata's other work, merely suggests conventional function but would not be used as a serious storage container. However, Kuramata declared that, since childhood, he had loved drawers "full of toys and spinning tops and colored cards. . . untidy drawers." And the chest here might indeed house these items, but little else. Another version of the chest—"Furniture in Irregular Forms Side 1"—undulated from front to back, rather than side to side. Other Kuramata furniture—chairs, tables, and a large number of other chests—also explores the use of drawers and reveals his self-declared obsession for them (see p. 194). Kuramata's work owes much of its unique nature to his having had one foot in the East and the other in the West; he was active in this native Japan, collaborated with Ettore Sottsass in Milan, and traveled widely and frequently worldwide. Kuramata left behind a diverse and healthy body of work when he died at age 57.

Date: 1970. Materials: Lacquered board. Manufacturer: Cappellini International Interiors, Arosio, Italy. Photograph: Courtesy Cappellini.

Events of 1970:

Exposition universelle opened, Osaka, Japan (64,000,000 visitors).

"Boby" taboret cart was designed by Cesare "Joe" Colombo for Bieffeplast—popular with architects and artists and still in production today.

"Landscapes for Living," futuristic environments by Danish designer Verner Panton, were presented at the "Visiona" exhibition at the Cologne furniture fair, Germany.

Premiers: The videodisk, by Decca of Telefunken; microprocessor, by Gilbert Hyatt, U.S. (made and sold by the Intel Corporation from 1971); International Design Zentrum, Berlin; Ferrari "512 S Modulo" auto; ITC (International Typeface Corporation), by Burns, Lubalin, and Rondthaler, New York; World Trade Center, New York; childproof safety tops for containers; female jockey (Diane Crump) in the Kentucky Derby; Amtrak, America's national railroad; subway in Mexico City; legal divorce in Italy.

Literature: two feminist landmarks—Germaine Greer's *The Female Eunuch* and Kate Millett's *Sexual Politics: A Surprising Examination of Society's Most Arbitrary Folly.*

Theater: Sondheim's musical "Company," on Broadway, New York.

Art: Robert Smithson made land into art with pioneering earthworks. Artist Mark Rothko's suicide touched off an ugly law suit between his daughter and his dealer, to become the costliest and longest in art history.

Thor Heyerdahl sailed a papyrus raft from North Africa to the West Indies, proving that early transatlantic crossing was possible.

Chile became the first nation to elect an avowed Marxist, Dr. Salvador Allende Gossens, as a head of state.

Jimi Hendrix and Janis Joplin died from illegal drugs. Charles de Gaulle and Yukio Mishima also died.

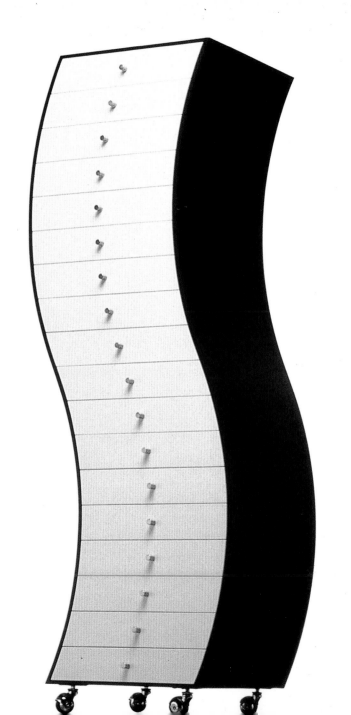

Richard Sapper (b. 1932) studied engineering in his native Munich, Germany, before he worked as a designer for a couple of years at Daimler Benz in Stuttgart. He subsequently, in 1958, settled in Milan where he eventually began a distinguished collaboration with Marco Zanuso (see p. 140). The "Tizio" lamp, Sapper's début effort working alone, illustrates the level of engineering knowledge he has brought to the design of consumer products. Possibly the most famous lamp in the world, the "Tizio," Italian for "what's-his-name," may echo characteristics of the Anglepoise- or Luxo-type lamp but is quite different. Sapper's lamp, like Buquet's of 1927 (see p. 66), has a counterbalanced arm that eliminates the necessity of springs but, unlike Buquet's, does not rotate on a horizontal plane. The fixture illustrates Sapper's credo: form is the servant of technology and materials. Thus, he invented a lamp that incorporated low-voltage halogen electrical wires thin enough for concealment inside the skinny structural elements, a base to house a heavy transformer, and a small reflector to hold a powerful, diminutive, and almost weightless bulb. The arms articulate without effort about snap joints, accented by red dots, and can be placed in almost infinite positions. The lamp—delicate, like a Calder mobile, and cheap to make—was first available in black only.

Design: 1972. Materials: Aluminum, ABS resin (base with a transformer), halogen bulb. Manufacturer: Artemide, Milan, Italy. Photograph: Courtesy Artemide.

Events of 1971:

The U.S. *Mariner 9* became the first spacecraft to orbit another planet, Mars. It was followed 6 and 7 months later by the Soviet *Mars 2* and *Mars 3,* including a soft-landing capsule.

Taking their forms from comic books and other Pop images, the "Cactus" coat rack by Guido Drocco and Franco Mello and the "Pratone" designed for lying and sitting on by Gruppo Strum were made possible by a new foam.

Premiers: The computer scanner, U.S.; "Corail" high-speed train, by SNCF, France; Lamborghini "LP 500 Countach" racing auto; Renault "R5" auto, by Michel Boué, France; Transac, automatic money distributor, by Bull, France; ban on televised cigarette advertisements, U.S.; floppy disc for computer storage, by IBM; direct dialing between New York and London; Walt Disney World, Orlando, FL, U.S.

Films: Friedkin's *The French Connection*, Parks's *Shaft* with Richard Roundtree, Kubrick's *A Clockwork Orange*, Visconti's *Death in Venice* with Dirk Bogarde, *Dirty Harry* with Clint Eastwood, *Fists of Fury* with Bruce Lee.

Literature: Industrial designer Papanek's *Design for the Real World* and Forsyth's *The Day of the Jackal*. Jean-Paul Sartre was indicted for libeling the police, Paris.

Music: Led Zeppelin's song "Stairway to Heaven," Bernstein's "Mass," the musical "Jesus Christ Superstar."

Conceptual art became a fad, U.S.

The Getty Museum bought Titian's *The Death of Actaeon* for $4,032,000, second highest price to date for a work of art.

The microprocessor or "computer chip" spawned a third generation of "microcomputers" (about the size of a TV set).

Taking place amid famine and political repression, the Shah commemorated the 2,500th anniversary of the Persian empire of Cyrus the Great with a $200,000,000 party.

"Coco" Chanel, Nikita Khrushchev, and Igor Stravinsky died.

The Polaroid Corporation, established in 1937 with the assets of Dr. Edwin Herbert Land (1909–91), applied the research of George Wheelwright III into light polarization (thus the name Polaroid) to microscopes, lamps, sunglasses, and cameras. Land who ultimately held 533 patents conceived the idea of a one-step integrated camera-and-film system in 1943. His stretching the possibilities of state-of-the-art technology resulted in the world's first example, the "No. 95" of 1948. The initial series produced only black-and-white prints; a color version was introduced in 1963. Henry Dreyfuss (1904–72) who had been active in his own industrial-design office in New York since 1929 was, from the beginning, part of the design team on the "SX-70" model. Working with Polaroid technicians who were conducting continuous research into chemistry, materials, electronics, optics, and processes, Dreyfuss's staff changed the camera's configuration almost daily. A reduction in the size of the device was made possible through two innovations. A power source, built into the film pack, eliminated the need for an area for batteries, and a through-the-lens viewing system allowed for the placement of the view finder almost completely inside the body. Chromium plating made the inexpensive, lightweight plastic housing rigid. Leather panels imbued the camera with a sense of value. With his terminally ill wife, Henry Dreyfuss committed suicide shortly after this project was completed.

Date: 1972. Materials: Chromium-plated plastic, leather, metal, glass. Manufacturer: Polaroid Corporation, Cambridge, MA, U.S. Photograph: The George Eastman House, Syracuse, NY, U.S.

Events of 1972:

"Italy: The New Domestic Landscape," the landmark exhibition of contemporary Italian design, was organized by Emilio Ambasz at The Museum of Modern Art, New York.

"Living Model" exhibit by Olivier Morgue spoke for "living landscapes" and was realized as "democratic" living spaces with a futuristic vision, at the "Visiona" stand by Bayer Leverkusen at the Cologne furniture fair, Germany.

Pictograms were designed by Otl Aicher for the Munich Olympic Games, first since World War II in Germany, in which U.S. swimmer Mark Spitz took a record 7 gold medals and Soviet gymnast Olga Korbut won 3 golds and a silver. The PLO group Black September, all of which were eventually killed, murdered 11 Israeli Olympic participants; one policeman died.

Premiers: The Nike sports shoe; European wide-bodied airplane ("Airbus A300"); commercially successful video game, "Pong," by Atari; electronic pocket calculator, by Texas Instruments, U.S.; Pentagram industrial design studio, London; Honda "Civic" auto, Japan; enkephalin painkiller, by Dr. John Hughes, U.S.

Films: Buñuel's *The Discreet Charm of the Bourgeoisie*, Coppola's *Godfather*, M. Camus's *Con el*.

Jamaican reggae singer Bob Marley became an influential music star.

Literature: Pirsig's *Zen and the Art of Motorcycle Maintenance*, Woodward and Bernstein's *All the President's Men*, Comfort's *The Joy of Sex*, Venturi and Scott Brown's *Learning from Las Vegas*, which discussed Pop architecture.

A break-in at the Watergate hotel-office building launched investigations into Richard Nixon's presidency, resulting in his resignation in 1974, Washington, D.C.

Maurice Chevalier and Hans Scharoun died.

In 1971, Carl Sontheimer (1914–98), an M.I.T.-trained physicist and a skillful cook, founded the Cuisinarts Co. (later Cuisinart) to import high-quality cooking wares. Visiting a trade show in France in 1971, he and his wife Shirley saw a professional food-preparation machine, known as the Robot-Coup when first developed by chef Pierre Verdun in 1963, which, by 1971, for the home market, had been perfected into the more compact Magimix. Rather than continuing with his importing business, Sontheimer purchased prototypes of Verdun's device, acquired the U.S. distribution rights, lengthened the feeding tube, improved the cutting blade and disks, added mandatory American safety features, and named it the "Food Processor." In January 1973, the machine was introduced at the National Housewares Exposition in Chicago. Priced at $140, four times the price of a blender, the machine was received with little interest. Undeterred, Son-theimer further improved the disks and blades in 1974 and the next year demonstrated his machine to high-profile American chefs whose enthusiasm infected food journalists. Sales increased appreciably, from a few food processors a month to hundreds a month. During the 1977 Christmas season when supplies of Cuisinarts had become depleted, U.S. retailers began selling empty boxes with the promise of later redemption. The Sondtheimers sold the firm to an investor group in 1988 which in turn sold it to the Conair Corporation in 1989.

Date: 1972–73. Materials: Plastics, steel, other metals. Manufacturer: Cuisinart, Stamford, CT, U.S. Photograph: Courtesy Cuisinart.

Events of 1973:

An oil-tanker, the first to be grounded off France, spilt crude oil and damaged the environment.

The Sydney Opera House by Danish architect Jörn Utzon opened after 16 years of construction, Australia. Chicago's Sears Tower at 1,454 feet became the world's tallest building.

Global Tools radical-design group was founded, Milan.

Premiers: The supermarket bar codes on goods; airport security for the prevention of terrorism; commercially available facsimile machine (fax or télécopier); "Divisumma" calculators, by Mario Bellini for Olivetti, Italy; trademark for Cotton Incorporated (National Cotton Council), by Landor Associates; Bell Telephone trademark, by Saul Bass, U.S.; interior of the *Concorde* supersonic airplane, by Raymond Loewy; laboratory housing in outer space, "Skylab," launched by NASA; rabies vaccine that required five injections (replacing 14 to 21 injections).

Film: Bertolucci's *Last Tango in Paris*, Bergman's *Scenes from a Marriage*, Friedkin's *The Exorcist*.

Music: Marley's song "I Shot the Sheriff," Elton John's LP "Goodbye Yellow Brick Road." Pink Floyd's bleak LP "Dark Side of the Moon" was to spend 741 weeks on U.S. music charts.

World-wide oil crisis was precipitated by OPEC's decision to greatly increase oil prices.

After 20 years of dumping methyl mercury into Yatsushiro Bay, the Chisso Corp. paid $600,000,000 in damages to 2,000 human victims, Minamata, Japan.

Juan Perón returned to Argentina and was elected president, and his nightclub-dancer wife became vice-president, Argentina.

Pro-abortion Supreme Court decision released the world's most militant anti-abortion movement, U.S.

W.H. Auden, Noel Coward, Pablo Neruda, Pablo Picasso, and Edward Evans-Pritchard died.

Like numerous other designers who became prominent in post-war Italian design, Mario Bellini (b. 1935) studied architecture at the Politecnico in Milan. Since opening his own office in 1962, he has become an highly active designer and architect in a large Milanese studio with prestigious clients including Cassina, Rosenthal, Ideal Standard, Vitra, and Renault. Bellini was always interested in not only how a machine looks overall but also its appearance in profile. For example, in 1963, Bellini first incorporated the wedge shape and slanting keyboard in a magnetic character encoder for Olivetti, a longtime client. For the machine shown here, he and his brother Dario (b. 1937) sliced off a corner of a rectangular box shape to serve both form and function: an interesting profile and a convenient tilt to better facilitate use. The busy but well-organized fascia fulfilled the imperative of Japanese makers of electronic equipment of the 1970s to offer electronics that looked, but were not, the same as those used by professional sound technicians. This High-Tech trend, emanating from some architecture of the later 1960s, soon found its way into homes. Directed to the male market, sound equipment with black fascia were proudly and prominently placed on display by consumers. Most of the ersatz knobs, levers, dials, and illuminated buttons had little effect on regulating the sound quality which was already excellent.

Date: 1974. Materials: ABS resin, acrylic plastic, metal. Manufacturer: Nippon Gakki Company for Yamaha, both Hammatsu-City, Japan. Photograph: Courtesy Yamaha.

Events of 1974:

Soviet car Moskwitch was designed by Raymond Loewy, who thus become the first designer in the West to receive a Russian contract.

A trophy in solid 18k gold of the World Cup of Football was designed by Italian sculptor Silvio Gazzaniga to replace the original version that survived World War II and theft before being retired to Brazil in 1970.

Premiers: The puzzle cube (patented 1975), by Hungarian architect Ernö Rubik; black mannequin on the cover of a major fashion magazine (Beverly Johnson on American *Vogue*); Ford "Mustang" auto body, by Eugene Bordinat and others; Volkswagen "Golf" auto body, by Giorgietto Giugiaro of Milan; computer-data memory card, by Roland Moreno; Roissy-Charles de Gaulle airport, near Paris; combination of black-hole theory and thermodynamics, by Stephen Hawking, U.K.; Heimlich maneuver for choking victims, by thoracic surgeon Henry J. Heimlich.

Literature: Solzhenitsyn's *The Gulag Archipelago: 1918– 1956*, Stoppard's "Travesties," Benchley's *Jaws*.

West German sculptor Joseph Beuys became one of the most controversial artists of the time.

Soviet dancer Mikhail Baryshnikov defected to the West.

Computers with LSI (large-scale integrated) chips made them capable of performing advanced trigonometric and logarithmic functions with prices reduced from $400 to $100 in this year alone.

Linking man to ape, a 3,200,000-year-old hominid, named Lucy, was found, Ethiopia.

India detonated its first atomic bomb (15 kilotons), much smaller than the U.S. bomb dropped on Hiroshima in 1945.

"Duke" Ellington, Charles Lindberg, Darius Milhaud, Juan Perón, and Georges Pompidou died.

From 1958–65, Gaetano Pesce (b. 1939) studied architecture and design in Venice. In the late 1950s, he experimented with art and collaborated with avant-garde groups in Germany, France, and Italy. Pesce rejected the rationalism and smooth contours that were appearing in Italian design in the late 1960s in favor of an approach found in the Pop, kinetic, and conceptual art movements. His explorations into plastics began in 1962 and with furniture in 1968. The two were combined in the "UP" series of inexpensive chairs of 1969 by C&B Italia. The "UP 7," a flesh-colored giant foot created a stir in the design community. His *Manifesto on Elastic Architecture* (1965) expounded on what he considered to be the alienation between people and objects in the consumer culture. He created a number of pieces of furniture—apart from his "nihilistic," decaying work—that included the "Sit Down" club chair (shown here), sofa, and ottoman. Ever one to create anew, the "Sit Down" continued an idea he had earlier explored in polyester and resin. For the "Sit Down," a plywood box was built in an inverted chair shape. Quilted fabric lined the box. Polyurethane, poured into the envelope, expanded into a foam that penetrated the folds of the fabric and was filled up to the top. Each seat was similar, but slightly different. The centuries-old technique for constructing fabric-upholstered, spring-constructed furniture had been banished.

Date: 1975. Materials: Quilted polyester fabric, polyurethane foam. Manufacturer: Cassina S.p.A., Meda (MI), Italy. Photograph: Courtesy The Museum of Modern Art, New York. Gift of the manufacturer. Photograph © 1999 The Museum of Modern Art.

Events of 1975:

A Soviet spacecraft was sent to Venus and broadcast the first close-up images of the planet.

"Postmodern" as a term for new architecture, succeeding the "Modern Movement," was coined by Charles Jencks, an American architect-writer living in England.

Premiers: The disposable razor by Bic; digital watch; personal camcorder (video-tape camera), by Sony, Japan; Merrill Lynch "bull" logo, by King-Casey Designs; American Express logo, by Lippincott & Margulies, U.S.; Organization of Black Designers, organized by David H. Rice, Detroit, Michigan, U.S.; Microsoft Corporation, by former computer hobbyists William H. Gates and Paul Allen, U.S.; theory of fractile images, by Benoit Mandelbrot; video to promote a pop-single song, by Queen.

Film: Forman's *One Flew over the Cuckoo's Nest*, Franco's *Pascual Duarte*, Kubrick's *Barry Lyndon*.

Music: Babbitt's "Reflections," Springsteen's LP "Born to Run," musical play "A Chorus Line."

Literature: Clavell's *Shōgun* and Wilson's *Sociobiology: The New Synthesis*.

First personal computer ("Altair 8800," with windows, icons, a mouse by William Englebart, and a pointer) was demonstrated by Xerox, Palo Alto, California.

7,000-piece ceramic army of the first Chinese emperor Shih Huang-ti of the first century B.C. was found.

Portugal became the last to abandon African colonization.

An earthquake killed 4,700 and injured 5,000, Pakistan.

Georges Charpentier, Francisco Franco, Aristotle Onassis, Dmitri Shostakovich, and P.G. Woodhouse died.

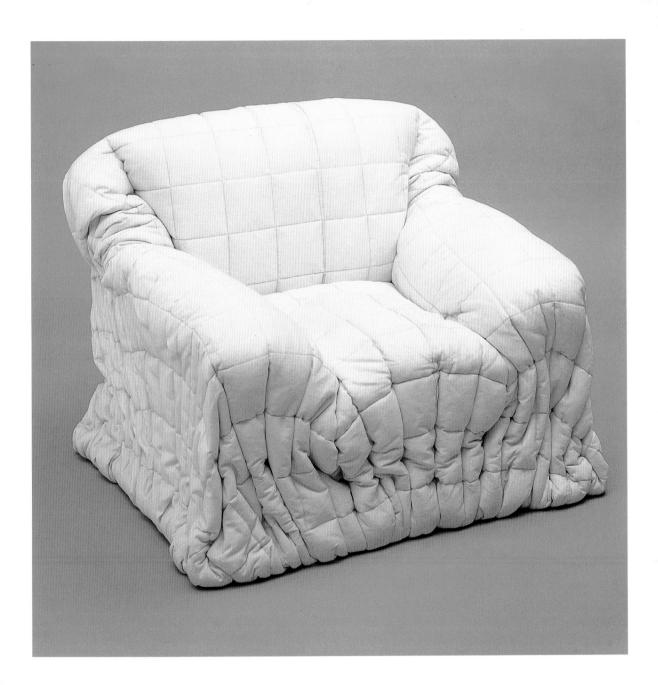

After studying history in America and living in Florence, Italy, for a short time, Michael S. Zane (b. 1948) returned to Boston, Massachusetts, where his father operated a lighting company and sheet-metal shop. Zane accidentally discovered a bicycle lock patented by Stan Kaplan, a bike-shop mechanic. After a brief partnership with Kaplan, Zane bought the rights to the lock in 1973 and agreed to pay royalties to Kaplan. Zane has reminisced about the 1971 design: "The original Kryptonite, of which only 50 were produced, looks like an Iron Age monstrosity when compared to the present design." Moving toward refinement, he sought advice from lock users, retailers, and college professors. His quest was spurred by a need to make money, and the solution drew upon the facilities provided by his father's metal shop. Finally, in 1977, Zane combined a patented "bent foot" with a tubular-steel cross bar, which lightened the weight by 675%. A sophisticated Ace II tubular key-lock cylinder and a patented disappearing hinge were added. When closed and locked, the hinge could not be pulled away from the cross bar. The no-fuss "Kryptonite" could be quickly and easily attached or removed and was completely vinyl-coated to protect the finish of a bike's frame. Further deterring theft, neither bolt cutters nor cable cutters could grip the flat cross-shank. Zane aptly named his lock after kryptonite, the only substance that can make Superman weak.

Date: 1976. Materials: Steel, vinyl. Manufacturer: KBL (today Kryptonite) Corp., Canton, MA, U.S. Photograph: Courtesy Michael S. Zane.

Events of 1976:

Studio Alchimia radical-design group was founded, Milan.

Cooper-Hewitt National Design Museum, New York, became an official entity in its own building, under the sponsorship of the Smithsonian Institution.

Premiers: The supercomputer ("Cray I"), producing 250,000,000 operations per second; toothbrush with an ergonomically bent handle ("Reach"), by Percy Hill and John Kreifeldt, sold by Johnson & Johnson, U.S.; "Accord" auto, by Honda, Japan; synthesis of a complete, functioning gene, U.S.; non-stop New York-to-Tokyo air service, by Pan Am; subway system, Washington, D.C.; Apple Computer, California.

The earliest bronze man-made object ever (of 3600 B.C.) was found, Thailand.

Films: Rocky with Sylvester Stallone, Pakula's All the President's Men, King Kong with Jessica Lange, Scorsese's Taxi Driver, The Omen with Gregory Peck.

Literature: Gaddis's JR, Marquez's The Autumn of the Patriarch, Haley's Roots, Vonnegut's Slapstick.

Photorealism gained popularity in painting.

Peugeot absorbed Citroën to become a single auto-manufacturing entity, France.

"Aerodynamics" as an auto-marketing philosophy was employed by Porsche in its advertising for the "924" model.

Exiled Soviet scientist in London claimed that the atomic blast of 1958 had killed hundreds in the U.S.S.R.

The U.S. celebrated its 200th anniversary.

30 Jewish activists were arrested in a protest, Moscow.

India began penalizing parents with more than two children.

Alexander Calder, Fritz Lang, Lily Pons, Man Ray, Carol Reed, Mao Zedong, and Luchino Visconti died.

"Plack" picnic set by Jean-Pierre Vitrac

From 1968 Jean-Pierre Vitrac (b. 1944) worked for the cosmetics firm Lancôme after graduating from the École des Arts Appliqués in Paris and studying sculpture. He specialized in packaging and in-store advertising before forming J.-P. Vitrac Design in Paris in 1974 to pursue both design and marketing projects with affiliates in Milan, New York and Tokyo. Housewares and other products included office furniture for Achilles, the "Lollipop" folding chair for Prodamco, a computer console for Agora, folding eyeglasses for K-Way, and a motor scooter for Bunny Courses. In addition to managing a highly active design firm, he wrote a book explaining how the quality of industrial design can help boost sales. Vitrac's interest in the use of new materials and technology was exploited in "Futura," a set of plastic-molded luggage of 1983 for Superior SA. His picnic set for Diam Polystrène which was marketed by Vitrac's firm Co. & Co. was a commercial failure yet it has remained his favorite design and canonical of the period. The difficult-to-solve assignment mandated that the five utensils necessary for use during a meal be incorporated into a single unit. The configuration of a knife, fork, spoon, cup, and plate was produced in polystyrene, a non-biodegradable material appropriate for use with food. In the 1970s, disposable plastics were not an ecological concern, and recycling was not in the public consciousness. Children particularly liked the picnic set because each item could be ripped or audibly broken away with some gusto as if they were destroying an object.

Date: 1977. Material: Pre-colored injection-molded polystyrene. Manufacturer: Diam Polystrène, France (from 1979). Photograph: Courtesy Jean-Pierre Vitrac, Paris.

Events of 1977:

Centre Georges Pompidou art museum, designed by Renzo Piano and Richard Rogers, opened, Paris.

Premiers: "Apple II" personal computer; fiber optics for telephone lines; nuclear reactors using enriched uranium for the production of electricity, by Fessenheim; "Microvision" miniaturized television set, by Sinclair Radionics, St. Ives, England; portable microcomputer, by Hewlett-Packard, U.S.; U.S. ban on Red Dye No. 2 that was found to cause cancer; commercial America-to-Europe supersonic air flight by *Concorde*.

Film: Lucas's *Star Wars*, Allen's *Annie Hall*, Wajda's *Man of Marble*, the Tavianis' *Padre Padrone*, Innaurato's *Gemini*.

Music: Fleetwood Mac's LP "Rumours," Eagles's LP "Hotel California," Faberman's "War Cry."

Literature: Grass's *The Flounder* and Caputo's *Rumor of War*.

Writing was traced back to 10,000 B.C., and the planet Uranus's rings were discovered.

U.S. sales of the "Beetle" auto were discontinued.

UNESCO appealed for donations to rescue the crumbling 2,400-year-old Acropolis, Athens, Greece.

Two planes collided, and 574 passengers died on the runway in the worst aviation disaster to date, Canary Islands.

Nicaraguan president Somoza was charged with atrocities.

Black smokers (or 300° C hydrothermal vents), that support animal life, were discovered in the eastern Pacific Ocean.

Earliest-known life forms, 3.4-billion-year-old microcells, were found in South African rocks.

First free elections since 1936 were held, and dissident Dolores Imbarruri returned after 38 years in exile, Spain.

Financing of the neutron bomb, to kill people but not to destroy buildings, was approved, U.S.

Maria Callas, Wernher von Braun, Henri-Georges Clouzot, Elvis Presley, and Leopold Stokowski died.

Donald Wallance (1909–90) was born in New York City where he studied at New York University and the Design Laboratory in the 1940s. He was active in design for the war effort and, in 1949 with his wife Shula, set up a studio in Croton on Hudson, near New York City. He is best known for his stainless-steel flatware for Lauffer in the U.S., but he also designed for Pott in Germany. Wallance's "Design 10" plastic utensils were, as the name suggests, his 10th design for Lauffer. However, this synthetic flatware was not the first. David Mellor, a highly accomplished British designer, had designed plastic flatware in 1969. Mellor's cutlery for Cross Paperware was so successful with customers that, even though it was disposable, they retained, washed, and reused it. Numerous others also made distinguished plastic utensils during the 1970s including Sarvis's "Easy Day" by Kay Franck (of the same year as Wallance's pattern) in the durable, reusable material Melamine. For his "Design 10," the obsessive Wallance made wooden models first and plastic models next and conducted extensive tests with various plastics. He found that Lexan by General Electric satisfied his demands for durability, dishwasher safety, and stain resistance. Specially treated molds created an attractive matte finish that resisted scratching. A handsome form in a range of colors featured generous spoon bowls. Franck's very similar pattern is still sold today; Wallance's is not.

Date: 1978–79. Material: Lexan (polycarbonate). Manufacturer: H.E. Lauffer Company, New York (from 1981). Photograph: Wallance Collection, Cooper-Hewitt National Museum of Design, Smithsonian Institution.

Events of 1978:

Spain's first design magazine, *On*, was founded.

Examples in the "Distortion of Great Works" series, including the alteration of Gerrit Rietveld's "Zig Zag" and Marcel Breuer's "Wassily" chairs by Alessandro Mendini, questioned traditional design concepts and icons, Italy.

The AT&T Building (to 1982) designed by Philip Johnson provoked lively discussions in the architecture and intellectual communities, New York.

Premiers: The "test tube" baby (Louise Brown), U.K; European rocket (*Ariane*), by Kourou; Computer Museum, initiated by an effort to save MIT's "Whirlwind" computer, by Ken Olsen and Bob Everett, Boston; ultrasound; legal abortion, Italy; Pinyin, official transliteration system for Chinese names; Ford "LTD II" auto, one of many "gas guzzlers" of the period; first non-Italian pope since 1523 (John Paul II), Vatican.

Music: Rolling Stones "Some Girls," Penderecki's opera "Paradise Lost," Bee Gees's "Stayin' Alive."

Film: Fassbinder's *The Marriage of Maria Braun*; Coppola's *Apocalypse Now*; *Saturday Night Fever,* acknowledging the peak of disco music; Donner's *Superman*.

Literature: Murdock's *The Sea* and Jencks's *The Language of Postmodern Architecture*.

Theater: Webber and Rice's musical "Evita" and Pinter's play "Betrayal."

Amoco Cadiz, second tanker to ground off the French coast, spilled 250,000 tons of oil and killed wildlife, Brittany.

Sweden become the first nation to curb ozone-destroying aerosol sprays.

Afghanistan became communist with a revolutionary coup.

Former prime minister Aldo Moro was kidnapped and killed by the Red Brigades, Italy. The daughter of couturier Calvin Klein was kidnapped but recovered safely, New York.

The coffin of Charlie Chaplin was stolen from his grave.

In 1946 soon after the end of World War II, Tokyo Tsushin Kogyo Kabushikakaika (TTK) was founded by Akio Morita and Masaru Ibuka. In 1958, the corporation became known as Sony, a name so chosen to sound more accessible to Westerners. It introduced many of the first all-transistor electronics: the radio in 1958, TV in 1960, audio-tape recorder in 1961, and video-tape recorder for home use in 1963. Sony's 1978 invention, the now legendary "Walkman," was a whole new concept in electronics for the general consumer. Purportedly the device was conjured from the restless mind of Sony chairperson Morita while he was playing tennis. Calling on Sony's existing transistorizing technology, a unit that played audio cassettes or received radio signals was built small enough to fit in a person's pocket or hang on a trouser belt. Small headphones were the most difficult problem to solve. The original see-through model (shown here) is a far cry from later versions that featured stereo sound, deep bass, and an array of control knobs, dials, and buttons. The "Walkman," like so many other examples of innovative electronic equipment, was somewhat expensive when first produced. As a result of subsequent versions by a large number of other manufacturers, the "Walkman" today is inexpensive, and its proprietary name has become generic. Further developments in electronics made cassette players obsolete and have been replaced by CD and even higher-density-disk machines.

Date: 1979. Materials: Plastics, metal. Manufacturer: Sony, Tokyo, Japan. Photograph: Courtesy Sony Corporation of America.

Events of 1979:

Soviet astronauts Lyakhov and Ryomine ended a record 175-day space trip.

Causing unwarranted international concern, a runaway satellite (NASA's "Skylab" launched 1973) fell to Earth without incident.

Premiers: The laser printer, by IBM; British female prime minister (Margaret Thatcher); a pope's reception at the White House, Washington.

Music: Schwantner's "Aftertones of Infinity," Feinsmith's symphony "Isaiah," Pink Floyd's LP "The Wall."

Egypt undertook a major restoration of the Sphinx.

$2,000,000 was stolen from a Brink's truck, and 137 bank robberies were recorded in one month (August) in New York.

Long oppressed through pogroms, the exodus of Jews from the Soviet Union peaked; over 51,000 visas were issued.

273 died in the worst air disaster in U.S. history, Chicago.

Radiation was accidentally released from the nuclear reactor at Three Mile Island, Pennsylvania, U.S.

World economies were in poor condition resulting from decreased consumer spending, high inflation, elevated bank interest rates, soaring oil prices, and shrinking bank accounts.

General global unrest occurred, and much wide-spread anti-American sentiment was expressed.

The Shah of Iran was sent into exile. Charged with atrocities, Ugandan ruler Idi Amin Dada was driven out of his country. President Somoza left Nicaragua with the country in ruins. The madman Bokassa was overthrown as the self-imposed emperor of the Central African Republic. Convicted of murdering an opponent, Pakistani premier Ali Bhutto was hanged.

Lord Mountbatten was killed by Irish terrorist. Peggy Guggenheim, Pier Luigi Nervi, Mary Pickford, and Jean Renoir also died.

Arthur Fry (b. 1931) spent almost four decades as a technician at the 3M Company in Minneapolis. He began working there in 1953 while still a chemical-engineering student at the University of Minnesota. His most significant contribution was the "Post-it® Note," essentially an accidental invention, the case with so many other technological and scientific developments. Fry found out about an adhesive that another 3M scientist, Dr. Spence Silver (b. 1941), had developed. It was a substance that would stick lightly to many surfaces but remain tacky, even repositionable, after being removed. Fry realized that, when he applied Spence's adhesive to paper, it was perfect for a personal need. "Now, I had a bookmark that could stick to the page while exposing a part that wasn't sticky," he has recalled. Soon afterward, Fry discovered that a note he had written on his bookmark could be attached to a document he was sending to a fellow worker. "That's when I came to the very exciting realization that my sticky bookmark was actually a new way to communicate and organize information." Other office employees at 3M began asking Fry for samples. Hence, his bookmark was sold as the "Post-it® Note" by 3M from 1980 in the U.S. and from 1981 in Europe. Available today in a range of colors and sizes, with and without printed lines, the canary-yellow color is 3M's own registered trademark.

Date: *c.* 1979–80. Materials: Paper, a proprietary adhesive. Manufacturer: 3M Company, Minneapolis, MN, U.S. (from 1980). Photograph: © Gregory S. Krum, New York.

Events of 1980:

Ettore Sottsass with others founded Sottsass Associati, active in a wide range of design areas, and initiated the Anti-Design group Memphis, introduced at the Salone del Mobile in 1981, Milan. Another Italian group, Studio Alchimia became important internationally.

Alessi began producing household products designed by architects, Crusinallo di Omegna, Italy.

Premiers: The in-line skate; fax transmission, made possible by a digital-converting system; RU-486 "abortion" pill, by Roussel Uclaf, Paris; eradication of smallpox; MS-DOS, a program adapted by Microsoft and the choice by IBM to be the operating system of its first personal computer; "Biodesign" cameras, by Luigi Colani for Canon; "Monospace" auto, by Matra Espace, France; French rocket ("Ariane") launching.

Film: Lynch's *Elephant Man*, De Palma's *Dressed to Kill*, Fassbinder's *Berlin Alesanderplatz*, Armiñan's *The Nest*.

Literature: Burgess's *Earthly Powers*, Eco's *The Name of the Rose*, Frisch's *Man in the Holocene*, Golding's *Rites of Passage*, Murdock's *Nuns and Soldiers*.

Information pushed the origin of life on Earth back to 3,500,000 years.

Unmanned planetary probe, *Voyager 1*, launched in 1977, flew past Saturn and transmitted stunning pictures to Earth.

Solidarnose union led by Lech Walesa was created, and thus the Solidarity Movement, Poland.

A gold rush in the Amazon of Brazil yielded $50,000,000 in nuggets but resulted in bloodshed and damaged the rain forest.

Iraqi leader Saddam Hussein's invasion of Iran for oil-rich territory and access to a waterway produced great bloodshed and lasted for almost a decade.

Roland Barthes, Cecil Beaton, John Lennon, Jean Piaget, and Jean-Paul Sartre died.

Son of an Austrian architect, Ettore Sottsass
(b. 1917) began his own architectural career in
1947 in Milan, Italy, setting up The Studio. From
1958, he designed typewriters, other machines,
and office-furniture systems for Olivetti. In the
1960s, he became the father figure of the so-
called Anti-Design movement in Italy, aligned with
the hippie culture and student unrest in Europe
and America. His decidedly plain design work
for various firms reflected the radicalism of the
times. Always having had an affinity with young
designers akin to his own fundamental approach,
Sottsass contentedly functioned within groups
while retaining his own design personality. His
Anti-Design inclinations became stronger from
1979 when he and others established Studio
Alchimia. He then founded Sottsass Associati and
the next year, 1981, again with others, produced
the Memphis furniture and objects collection that
was intended to ridicule the bad taste of the mass
population. Augmenting the folly, the Memphis
furnishings were expensive. However, the advent
received unanticipated praise from the press. The
"Carlton" bookcase shown here, in Memphis's first
collection, is an extreme form covered with plastic
laminates in vulgar colors and patterns. Ever the
maverick, Sottsass dissolved Memphis in 1988
while it was still highly successful, declaring that it
was an idea whose time had passed.

Date: 1981. Materials: Wood, Abet Laminate plastic sheeting.
Manufacturer: Memphis, Milan, Italy. Photograph: Courtesy
Sottsass Associati S.r.l., Milan. Photo: Aldo Ballo.

Events of 1981:

Radio in a clear-plastic bag was designed and produced by
Daniel Weil, U.K.

"Cocktail" design group, exponents of the New German
Design, was founded by von Brevern and Mühlhaus, Berlin.

Premiers: Legal divorce in Spain; cloning of a mammal (3
mice), Switzerland; truly practical, reliable, and cheap microcomputer ("PC/AT"), by IBM; a solar-powered airplane flown
over the English Channel; voyage into space of the first
reusable vehicle (the *Columbia* "STS-1" shuttle); flight of the
"Stealth" bomber "F-117" (not announced publicly until
1988); CD (compact disk) recorder, by Philips, the
Netherlands; Pac-Man video game; TGV high-speed train,
France; fashion collection of Rei Kawakubo.

Pablo Picasso's painting *Guernica*, which recognized the
agony of the Spanish Civil War, was returned to Spain, under
terms of his will, from The Museum of Modern Art, New York.

Films: Hudson's *Chariots of Fire*, Spielberg's *Raiders of
the Lost Ark*, the Bartolomes' *Después de...* (After...), Szabó's
Mephisto, Fellini's *City of Women*, Betancor's *Days of Dawn*.

Literature: Rushdie's *Midnight Children* and Schyler's
Morning of the Poem.

Theater: Shaffer's "Amadeus" and Webber's "Cats."

The AIDS epidemic was recognized.

MTV became the first 24-hour music TV channel, U.S.

François Mitterand was elected the president of France.

Hosni Mubarak succeeded the assassinated Egyptian leader
Anwar Sadat.

A terrorist shot Pope John Paul II, Vatican, and a deranged
person shot U.S. president Ronald Reagan and others,
Washington.

Lady Diana Spencer and Prince Charles were married, U.K.

Abel Gance, Marcel Breuer, and Albert Speer died.

The personal computer appeared in infant forms in 1975 as a crude hobbyist version, the "Altair 8080" (see p. 172), and the more sophisticated "Apple I" (see pp. 180, 186). By the early 1980s, very few PCs could be found in offices and were practically unheard of in homes. A portable version was even rarer. For mass appeal, manufacturers working with industrial designers began focusing on making the operation of computers easy or, what has become a tiresome catch-phrase, "user friendly." Grid Systems' entrance into the portable-computer market extends back to 1979 when Bill Moggeridge (b. 1943) established I.D. Two, a California branch of his London-based design firm. The next year, Grid hired the Moggeridge team, known today as IDEO, "for an unusually close and ongoing relationship between the industrial design group and the rest of our [technical] team," according to Grid head John Ellenby. Moggeridge sought to produce a laptop that "in the portable state, . . . should be simple, hard, severe, technical and extremely robust. When opened, it should reveal a visually exciting and friendly interior. . . [Its] most interesting appearance would be achieved when the product was actually being used." Some of its features were unprecedented: an electroluminescent display window, a shallow keyboard, use of bubble memory, fold-out legs for a convenient tilt toward the user, and, in transit, a fold-down window that protected the keyboard.

Date: 1982. Materials: Plastic, metal, glass. Manufacturer: Grid Systems Corporation, Mountain View, CA. Photograph: Courtesy Bill Moggeridge, IDEO, Palo Alto, CA, U.S.

Events of 1982:

"Lost Furniture—More Beautiful Living" ("Möbel perdu") exhibition was held at the Museum for Art and Trade, Hamburg.

Premiers: Liposuction; "Watchman," by Sony; Halcion sleeping pill; California branch of frogdesign studio and work on the design of the Apple computer, by Hartmut Esslinger/ frogdesign; Domus Academy for design, Milan; experiments on the Minitel electronic telephone directory, France; superconductor, employing liquid nitrogen in its manufacture, U.S.; EPCOT (Experimental Prototype Community of Tomorrow) center, Orlando, Florida.

Films: The Tavianis' *The Night of Shooting Stars*, Fassbinder's *Lola*, Edwards's *Victor/Victoria*.

Music: McCartney and Wonder's LP "Ebony and Ivory" and Zwillich's "Symphony No. 1."

Literature: Naisbitt's *Megatrends*, Böll's *The Safety Net*, Márquez's *Chronicle of a Death Foretold*, Fuller's *A Soldier's Play*, Fugard's *Master Harold. . . and the Boys.*

Art: Basquiat's *Quality Meets for the Public* and Gilbert and George's *Cabbage World*. Neo-Expressionism was popular in Europe.

Manufacture of the DeLorean auto was halted, Belfast.

The French government nationalized many of it's banks and industries, counter to other countries' lowering inflation.

U.S. Air Force received the first delivery of the "F-117" (or Stealth Bomber) airplane (conceived in 1975, not deployed for combat until 1991 in the Persian War).

Argentina invaded the Falkland Islands, governed by the British for 149 years, and were defeated. Israel invaded Lebanon.

First hurricane in 23 years struck Hawaii.

Louis Aragon, Leonid Brezhnev, Princess Grace of Monaco, Jacques Tati, and King Vidor died.

By the 1970s, the Swiss watch industry was being smothered by competition from Asia, partially encouraged by low worker wages in the East. And the marketing, design, and technology of most watches had generally become outmoded. A consortium of Swiss watchmakers hired Nicholas G. Hayek (b. 1925) to survey its survival prospects. His survey showed that the public preferred Swiss watches, even at slightly higher prices over Asian models. Results further indicated that a competitive watch must be very well made, shock resistant, tough, waterproof, precise, analogical (with minute/second hands, not digital like Asian watches), and inexpensive, at $35. Encouraged by bank executives, Hayek bought the consortium with uncertain prospects. A solution for a cheap watch that married plastics with slimness came from a team lead by young engineer Jacques Müller. All elements were thoroughly considered; a normal watch's 151 parts were shrunk to 51. In 1981, a prototype appeared. By 1983, the merger of advanced manufacturing techniques, a sophisticated design, clever marketing, and what Hayek calls his "dreaming spirit" came together in a phenomenally successful product, modishly called the "Swatch." By 1984, Max Imgrüth's distinctive colors and patterns and soon motifs by well-known artists and others had altered the fashion-watch market forever.

Date: 1983. Materials: Metal, plastic, glass. Manufacturer: SMH Group (Suisse Microelectroniques et Horologes Compagnie S.A.), Switzerland; today Swatch Group, Biel-Bienne, Switzerland. Photograph: Courtesy Swatch Group.

Events of 1983:

The AIDS virus was identified by Dr. Luc Montagnier at the Institut Pasteur, Paris.

Stiletto (a.k.a. Frank Schreiner) designed the "Consumer's Rest" prototype chair, a supermarket cart recreated as a ready-made piece (serially produced from 1990).

"Cent ans de l'automobile française" (100 Years of the French Automobile) exhibition opened at the Grand Palais, Paris.

Premiers: The camcorder (recording video camera); Rubik Studio (designer of Rubik's cube puzzle), Budapest; Colorcore material, by Formica; a black person elected to the Academie Francaise.

Film: Bergman's *Fanny and Alexander*, Scorsese's *King of Comedy*, Kurys's *Entre nous*, Uribe's *La muerte de Mickel*, Attenborough's *Gandhi*. Popular Hollywood "Wild Style" films featured break dancing and rap music.

Literature: Thomas's *Swallow*, Coetzee's *The Life and Times of Michael K.*, Pulos's *The American Design Ethic: A History of Industrial Design*.

Music: Jackson's "Beat It," The Police's "Every Breath You Take," Powell's "Strand Settings: Darker."

A Degas painting sold for $3,740,000, a record for Impressionist art, London.

U.S. invades tiny, defenseless island of Grenada with great braggadocio.

254 U.S. Marines were killed in a Beirut barracks bombing.

Millions protested NATO deployment of U.S. missiles in West Germany, U.K., Stockholm, Rome, Paris, Brussels.

Ex-Gestapo officer Klaus Barbi who was a spy for the U.S. which helped him to escape Europe was arrested in Bolivia and returned to France for trial.

Luis Buñuel, George Cukor, R. Buckminster Fuller, Gloria Swanson, and Tennessee Williams died.

The computer mouse has imbued the personal computer with its accessible, user-friendly character. The handy device has made it possible for people to make the most universal of human gestures—pointing. The very essence of technological innovation, the mouse was born of research conducted by William C. Englebart (b. 1925). During World War II, he felt that it might be possible for the indications on a radar screen to be one of a "figural-visual" graphic nature around which he could glide. Ultimately arriving at a solution, his pointing device won out over others in a 1963–64 experimental study conducted by Bill English at Stanford University in California. Subsequently, in 1968 at Stanford during a multimedia presentation by a research group headed by Englebart, a multiple-viewing system, hypertext, and the mouse were demonstrated. But, due to a lack of vision in the Silicon Valley computer community, a workable personal computer did not appear until 1975 in the form of Jobs and Wozniak's "Apple I." Englebart had already applied for a patent for the pointer in 1967 which was granted in 1970 as the "X-Y Position Indicator for a Display System." Apple Computer assigned the refinement of Englebart's "indicator" to Hartmut Esslinger and his frogdesign staff who brought geometrical rather than ergonomic refinement to the little virtual creature (see p. 186).

Date: 1984. Materials: Plastics, metal. Manufacturer: Apple Computer, Inc., Cupertino, CA, U.S. Photograph: Courtesy Apple Computer, Inc.

Events of 1984:

Architect I.M. Pei designed the Pyramid (opened in 1989) of the Louvre museum, hated then but loved today by the French, Paris.

Café Coste designed by Philippe Starck became one of his first world-publicized commissions and resulted in the table and chair there being serially produced, Paris.

Premiers: The transplant of a baboon's heart into a human baby (Baby Fae) who died 15 days later, by Leonard L. Bailey, U.S.; transatlantic solo balloon flight, by Joe Kittinger, Maine to Italy; humans floating freely, untethered, in outer space, from the *Columbia* shuttle.

Music: Madonna's LP "Like a Virgin"; Boulez's "The Perfect Stranger"; Run-D.M.C.'s LP, the first rap album to sell "gold" (500,000 copies).

Film: Spielberg's *Indiana Jones and the Temple of Doom*, Forman's *Amadeus*, Lean's *A Passage to India*, Tabernier's *Un dimanche à la campagne*, G. Aragon's *La noche más hermosa*, Ishi's *The Crazy Family*.

Literature: Sillitoe's *Down from the Hill*, and Capra's *Turning Point*, to popularize the term "new age."

Theater: Webber and Stilgoe's "Starlight Express" and Sondheim and Lapine's "Sunday in the Park with George."

The U.K. agreed to return Hong Kong to China.

A French ship carrying uranium sank off Belgium.

Marchers protested private-school regulations, Paris.

Union Carbide gas leak killed about 2,100 people, India.

Morrie Futorian established the Stratford Company, which employed Ford-type assembly-line techniques to make cheap upholstered furniture, Tupelo, Mississippi, where 230 plants that make inexpensive furniture were located by 1998.

The Soviets boycotted the Olympic Games, Los Angeles.

Truman Capote, Indra Gandhi, François Truffaut, and Johnny Weissmuller died.

Kettle by Michael Graves

A practicing architect, Michael Graves (b. 1934) had begun spending at least half his time on product design by 1980. The decade that followed was marked by the appearance of expensive, high-style architect-designed products by Graves and others. Graves's kettle probably saw the greatest success. Its most distinctive feature—the playful form of a fledgling at the spout—emits a whistling sound when the water inside boils. (A whistling kettle, first shown at the 1922 Chicago housewares fair, was conceived by retired New York cookware salesperson, Joseph Block, while touring a German tea kettle factory.) On Graves's kettle, a blue arched pad protects the hand from the heat of the wire handle. A practical and visually pleasing design, the kettle's broad bottom encourages quick water heating, and its wide mouth facilitates easy cleaning. Even though the manufacturer may have at first felt Graves's fee of $75,000 to be extravagant, sales of 1,500,000 units since its introduction have proved the decision to have been a wise one. The "Juicy Salif" lemon press (see p. 192) by Philippe Starck is the only other Alessi product to have had sales comparable to the Graves kettle; however, the lemon press is far cheaper. Considered a so-called Postmodern object, the kettle became part of a complete tea set that included a sugar bowl, a creamer, a tray, and other items.

Date: 1985. Materials: Stainless steel (polished or, later, black silicon-resin coated), polyamide. Manufacturer: Alessi, Crusinillo di Omegna (VB), Italy. Photograph: Courtesy Alessi.

Events of 1985:

To be known for their kitsch approach to design, Pentagon and GINBANDE design studios were founded, Germany.

Premiers: A one-time-use disposable camera, by Fuji, Japan; cellular phones, U.S.; Nintendo video game; addition of a second to the calendar year; *Intramuros* design magazine, initiated by Chantal Hamaide, Paris; Transputer, allowing the management of parallel information processing, by Inmos, U.K.; an Arab in space.

Art: Purchase of Mantegna's 15th-century painting *Adoration of the Magi* by the Getty Museum for a record $12,000,000, Los Angeles; theft of Monet's *Impression, Sun Rising*, derivation of the painting term "Impressionism," and other paintings, Marmottan Museum, Paris.

Theater: Brook's staging of the 9 1/2-hour play "The Mahabharata"; worldwide broadcast of "Live Aid" rock concert, raising over $60,000,000 for African famine relief.

A U.K. meteorologist confirmed a hole in the ozone layer.

The *Atlantis* space shuttle made its maiden flight.

The only-known short story by the poet Lord Byron was discovered by his publisher, John Murray, London.

Mikhail Gorbachev unilaterally halted the deployment of medium-range missiles to Europe, U.S.S.R.

Switzerland became the first European country to mandate catalytic converters (that require lead-free fuel) for use in private autos.

The wreck of the ocean liner *Titanic* was found in the North Atlantic Ocean.

Terrorism was escalated sharply worldwide, including capture of a Rome-bound TWA plane and the *Achille Lauro* ship.

Laura Ashley, Heinrich Böll, Marc Chagall, Jean Dubuffet, and Simone Signoret died.

Norman Foster (b. 1935) studied architecture in Britain and the U.S. He set up a practice in London in 1963 and shortly after formed Team 4 with his wife Wendy Foster and Su and Richard Rogers. In the late 1960s, Foster, Rogers, and others began experimenting in a style which came to be known as High-Tech, an architecture partially based on the explicit exposure of structural elements and various operational details. In fact, the inside-out Centre Pompidou (1971–77) in Paris by Renzo Piano and Rogers who had already left Foster became the expression's apotheosis. The "Nomos" furniture system by Foster and Partners continued the themes of structural nudity and the adjustability of interior spaces to changing functions. All elements of the "Nomos" are on view. The system sought to provide a furniture range flexible enough to satisfy the demands of the office of the day. For use in homes as well, the presumption was that a table for an office differs little from one for a residence. The extensive inventory of interchangeable parts was to be assembled, arguably by the user, for various table and desk bases to hold any number of attached lighting fixtures, upper shelves, storage units, and other components including round, square, or rectangular glass or solid tops, some tiltable. It was hoped by the designers and manufacturer that standardization would encourage manufacture worldwide.

Date: 1986. Materials: Chromium-plated or powder-painted steel fused to aluminum, glass, rubber. Manufacturer: Tecno S.p.A., Milan, Italy. Photograph: Courtesy Tecno S.p.A.

Events of 1986:

Mies van der Rohe's German pavilion of 1929 for a Barcelona exposition was reconstructed. Lloyds Building was designed by Richard Rogers, London.

"The Machine Age in America, 1918–1941" was held at the Brooklyn (New York) Museum of Art, U.S.

"How High the Moon" armchair by Shiro Kuramata for Vitra in pierced metal has since become a famous 1980s emblem of symbolic, though not particularly functional, furniture.

Premiers: The global non-stop, non-refueling flight, flown by Jeana Yeager in a lightweight plane by Dick Rutan; "Windows," a computer program using icons on the screen (originally developed by Apple), by Microsoft, U.S.; practical software ("Alias") that allowed designers to manipulate form and color in real time and link up to traditional CAD systems, by Alias Research, U.S.; "Rafale," a prototype jet fighter airplane, France; U.S. poet laureate; AZT pill treatment for AIDS; Spain and Portugal as members of the European Common Market; state of superconductivity, the temperature at which a material loses electrical resistance, by Müller and Bednorz, Switzerland.

Andrew Lloyd Webber, whose musical "Phantom of the Opera" opened in London, became the most successful composer in theatrical history.

The *Challenger* space shuttle exploded, killing 7, over Cape Canaveral, FL, U.S.

Ex-U.N. secretary general Kurt Waldheim who lied about his Nazi past was elected the president of Austria.

France and Britain agreed to the construction of a tunnel under the English Channel.

A nuclear power plant melted down, Chernobyl, Russia.

Dictator Claude "Baby Doc" Duvalier was deposed, Haiti.

World population exceeded 5,000,000,000 inhabitants.

Simone de Beauvoir, Keith Haring, and Raymond Loewy died.

In 1975, Steven Jobs (b. 1955) and his friend Stephen Wozniak secured $20,000 worth of parts, sold some of their possessions for $1,350, and used the money to make the first ever practical, sophisticated personal computer. The machine was sold as the "Apple I" in 1976, and Jobs, Wozniak, and Ron Wayne founded Apple Computer. They followed with the "Apple II" that sold for $1,298 and was the first computer on the market with a color monitor and separate keyboard. Various models and improvements followed suit, and the firm became one of the fastest growing enterprises in America. In 1982, Jobs conducted a worldwide search for a consultancy to initiate Apple's official design program. Among 80 mostly European contenders, Hartmut Esslinger and his German firm, frogdesign, were the final choice. In 1982, more than 400,000 units of Esslinger's first assignment, the "Apple IIc," were sold. Being paid an annual fee of $2,000,000, Esslinger settled in California, initially working as a consultant-designer to Apple and also to an auto manufacturer. His design for the adorable "Apple SE" of 1985 (introduced in 1987) was distinctive in a sea of boring plastic boxes by other manufacturers of the time and more or less established beige as a standard color for computers.

Date: 1985–87. Materials: Plastic, metal, glass. Manufacturer: Apple Computer, Inc., Cupertino, CA. Photograph: Courtesy Apple Computer, Inc.

Events of 1987:

"Tabula Rasa" pull-out table by GINBANDE for Vitra, with its fantastical extension feature, questioned function and expounded on the table as a meeting place, Germany.

The French "Airbus A320" airplane made its maiden flight.

Artists petitioned the Pope to halt the "desecrating" restoration of Michelangelo's Sistine Chapel ceiling, Vatican.

Sotheby's auction house sold Vincent van Gogh's painting *Irises* for a record $53,900,000, New York.

Music: Jackson's LP "Bad," Prince's "Sign 'O the Times," Albert's "Anthem and Processionals," U2's LP "The Joshua Tree," Sondheim's musical play "Into the Woods."

Films: Bertolucci's *The Last Emperor*, Lyne's *Fatal Attraction*, De Palma's *The Untouchables*, Kurosawa's *Ran*, Kubrick's *Full Metal Jacket*, Russell's *Gothic*.

Dow Jones industrial average fell 508 points (22.6%), the greatest to date and twice that of the 1929 Crash.

Oprah Winfrey's talk show was first aired soon becoming the most successful in America and was also shown worldwide.

Chun Doo Hwan's electoral manipulations caused the largest protest movement in South Korea's history.

Freedom for 3 of 23 hostages was won by Church of England envoy Terry Waite who was himself eventually confiscated by Hizballah guerrillas, Lebanon.

In a marathon trial, more than 300 members of the Sicilian mafia were sentenced for a range of crimes, Palermo.

Klaus Barbie was sentenced to life imprisonment for Nazi war crimes, France. Rudolf Hess, a Hitler confidant and a prisoner since 1941, hanged himself in Spandau Prison, Berlin. Fred Astaire and Jascha Heifetz also died.

Advanced artistic ideas and liberal political and economic pursuits as well as a healthy design industry in Spain had been thwarted by the tight grip of Generalissimo Francisco Franco's repressive government. Even though social discontent had long been prevalent, constraints were not relaxed until the death of Franco in 1975. From 1978, Spain began experiencing an explosion of free expression not seen since the 1930s. Some designers worked in an informal, if renegade, manner, realizing designs of small-production furniture and bars, which have always been popular with Spaniards. A separate faction included architects such as Oscar Tusquets Blanca and Jorge Pensí (b. 1946); they worked on a different, more formal and industrial, level. Aided by Spanish as their native language, Pensí, his friend Alberto Liévore, and others had immigrated to progressive Barcelona from politically stringent Buenos Aires. The design industry, previously ignored by the government, was in the 1980s being embraced for the national good. Pensí's "Toledo" chair shown here is emblematic of the work of this time with "typically Spanish" references, such as to city names and the bull fight. Furniture designs were then finding mass manufacturers and exporters. The "Toledo," included in Knoll's line in America, encouraged the development of cast-aluminum furniture in Spain and redefined the language of furnishings for use outdoors.

Date: 1988. Material: Cast aluminum. Manufacturer: Amat S.A., Barcelona, Spain. Photograph: Courtesy Jorge Pensí.

Events of 1988:

Elle Décoration magazine began publication, Paris.

Javier Mariscal designed the "Cobi" mascot for the Barcelona Olympics.

Premiers: The Worldwide Web, Washington, D.C.; Prozac, a pharmaceutical for the treatment of depression; plutonium-powered heart pacemaker; U.S. advertising on Soviet TV; Olympus "LSD 1000" compact camera, Japan; a female leader in a Muslim country (Benazir Bhutto), Pakistan.

Film: Almodóvar's *Women on the Verge of a Nervous Breakdown*, Malle's *Au revoir les enfants*, Levinson's *Rain Man*.

Literature: Stephen Hawking's *A Brief History of Time* and Peter Dexter's *Paris Trout*.

More than 1,000,000 fax machines were sold in the U.S. with sales in Europe and Japan not far behind; Japan was the leading manufacturer and user.

Work began on the "Chunnel," the English Channel underground tunnel between France and England.

Carbon-14 dating concluded that the shroud of Turin, purported to be the burial cloth of Jesus, was a fake.

Pan Am flight 104 was bombed by Lybian terrorists, killing 270 passengers, over Lockerbie, Scotland.

Apple Computer, the originator of PC icons, sued Microsoft for infringing on its copyright but later lost.

First world AIDS conference, among 148 nations, was held.

Depletion of the ozone layer was globally acknowledged.

British writer Salman Rushdie was condemned to death by the Ayatollah Khomeini for his book *The Satanic Verses*.

Premier Gorbachev withdrew Soviet troops from Afghanistan.

Longest undersea tunnel (33.6 mi.) was constructed, for the Seikan Railroad, Japan.

Frederick Ashton, Louise Nevelson, Isamu Noguchi, and Thomas Sopwith died.

Alberto Meda (b. 1945) began his career as an engineer working for a time at several industrial firms. While working for the plastics firm Kartell (see p. 142) in 1973, he initiated research into the use of polyurethane. Paolo Rizzatto (b. 1941), who studied architecture to 1965, has designed lighting for Arteluce and furniture for Busnelli and Molteni while active also in architecture. From their own separate design studios, Meda and Rizzatto have collaborated on a number of projects including lighting. They designed the "Berenice" table lamp of 1984 and the "Titania" ambient-lighting fixture for Luceplan of Milan, Italy. When their "Titania" is viewed straight on, even closely, 24 blades screen the eyes from the glare of the bulb. A set of translucent polycarbonate blades, or filters, in violet, yellow, blue, red, and green, is provided with each lamp. The user can combine the colors at will. The body is offered in a natural or black-anodized aluminum finish. In the ceiling version, the lamp hangs from two whisper-thin nylon strands. The strands are threaded through the ceiling-bracket support. The height of the lamp is adjusted by a cast-aluminum counterweight attached to the ends of the nylon strands. The arrangement also permits the lamp to be angled. The "Titania" combines technological innovation and a certain sprightly nonchalance, although the designers claim more serious intent.

Date: 1989. Materials: Stamped and cast aluminum, polycarbonate, nylon. Manufacturer: Luceplan, Milan, Italy. Photograph: Courtesy Luceplan.

Events of 1989:

Two museums were established: Design Museum, by Terence Conran and others, Butler's Wharf, Docklands, London; and the Vitra Design Museum, by Alexander von Vegesack and Rolf Felbaum, Weil am Rhine, Germany, in a building designed by Frank O. Gehry.

Premiers: The complex reflector headlight, on the Honda "Accord" auto; worldwide ivory-trade ban; Time Warner Inc. (from Time Inc. and Warner Communications); insurance policy against computer-virus losses, U.S.

Music: Miami Sound Machine's LP "Let It Loose," Zwilich's "Symphony for Winds," Leonard and Ciccone's song "Like a Prayer," Rolling Stones's "Steel Wheels," Boubil and Schönberg's musical "Miss Saigon."

Film: Tan's *Joy Luck Club*, Sheridan's *My Left Foot*, Branagh's *Henry V*, Stone's *Born on the Fourth of July*, Greenaway's *The Cook, The Thief, His Wife and Her Lover*, Almodóvar's *Tie Me Up, Tie Me Down*, Burton's *Batman*.

Literature: Schama's *Citizens: A Chronicle of the Revolution*, King's *The Dark Half*, Michener's *Caribbean*.

Burma was renamed Myanmar, and its capital Rangoon became Yangon.

The oil tanker *Exxon Valdez* was grounded, spilling 11,000,000 gallons of crude oil into pristine waters off Alaska.

U.S. invaded Panama and captured General Manuel Noriega who had allowed cocaine shipments into America.

A political protest in Tiananmen Square sparked the biggest revolt and fracas since the revolution, Bejing, China.

The Berlin wall was torn down, signifying the end of the Cold War between the U.S.S.R. and the West.

Salvador Dalí, Daphne du Maurier, Ferdinand Marcos, and Laurence Olivier died.

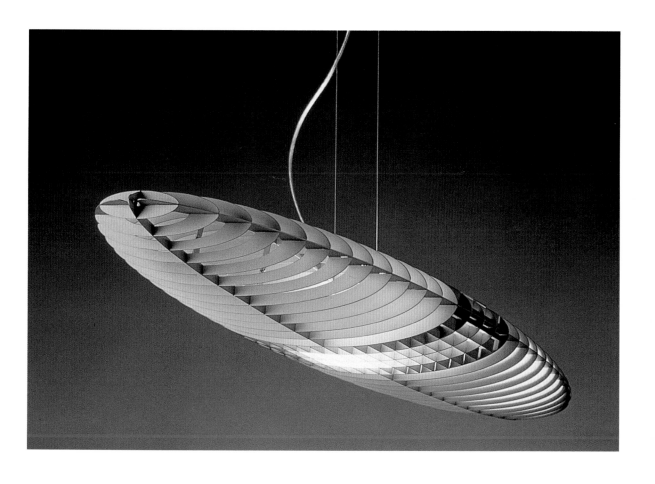

Philippe Starck (b. 1949) produced his first piece of furniture—the "Spanish Chair"—in 1968 but, as an independent designer, was largely undistinguished until the commission of a room in the Élysée Place of 1983–84 and the Café Coste in 1984, both in Paris. Backed by the publicity provided by these projects, Starck seized the moment and parlayed himself into a darling of the press through his unorthodox comments and behavior. By 1990, copies of the furniture for the Café Coste had sold in the hundreds of thousands. His clients for household products and furniture have included Driade, Disform, Thompson, and Sasaki. The range of objects has encompassed a water bottle for Vittel, a toothbrush for Fluocaril, and even pasta. Claiming an influence from science-fiction writer Philip K. Dick, Starck has designed buildings as well. Associated with the manufacturer Alessi since 1988, Alberto Alessi has said: "He is the living example of my dream: …always highly charged with innovation toward manufacturing and trade." Commissioned by Alessi to design a tray, the unpredictable Starck presented him with the "Juicy Salif" that was to become one of Alessi's best-selling products. The press for lemons may have more iconographic than functional value—a trend, as the public has become more appreciative of élitist design, toward displaying cult objects as sculpture.

Date: 1990–91. Materials: Polished cast aluminum (the initial version) or P.T.F.E.-treated to produced an anthracite color (another, subsequent version), polyamide feet. Manufacturer: Alessi, Crusinillo di Omegna (VB), Italy. Photograph: Courtesy Alessi.

Events of 1990:

Divided for 45 years, Germany became one nation again.

Iraqi forces invaded small Kuwait and then sacked it, setting off the first multi-national war in the post-Cold War era.

The U.S. *Magellan* spacecraft transmitted photographs of Venus showing a clear resolution within 30 meters.

Philippe Starck's toothbrush for Fluocaril, France, may have been the first high-style "designer" toothbrush.

Braun succumbed to competitive pressures from the Japanese and stopped producing hi-fi equipment, Germany.

Premiers: A McDonald's fast-food restaurant in Moscow; Hubble outer-space telescope (was optically flawed but repaired in 1993); gene therapy to repair or alter a body's cells, for a 4-year-old child, U.S.; Yamaha "Morpho II" experimental motorbike with organic styling, Japan; International Furniture Fair, New York; longest-serving British prime minister (Margaret Thatcher); ban of cigarette smoking on domestic U.S. air flights; performance of "the three tenors" (Domingo, Carreras, and Pavarotti), Rome.

Music: Prince's "Nothing Compares 2U," Living Color's "Time's Up," Powell's "Duplicates."

Film: Costner's *Dances with Wolves*, Besson's *La femme Nikita*, Salvatore's *Mediterraneo*, Zeffirelli's *Hamlet*.

Literature: Paglia's *Sexual Personae* and Walcott's *Omeros*.

Retrospective of photographer Robert Mapplethorpe caused the biggest art controversy in American history, Cincinnati.

Leftist priest Jean-Claude Duvalier became president of Haiti.

Sandinista dictator Anastasio Somoza was voted out, replaced by Violeta Barrios de Chamorro, Nicaragua.

Rabbi Meir Kahane was assassinated by an American of Egyptian descent, New York. Leonard Bernstein, Capucine, Erté, Greta Garbo, Armand Hammer, Rex Harrison, and Mary Martin also died.

Dutch designer Tejo Remy (b. 1960) studied 3-dimensional design at the Hogeschool voor de Kunsten (college of fine arts) in Utrecht. It was there for his final exam that he produced various furnishings, including the chest of drawers "You Can Lay Down Your Dreams." Drawing on the story of Robinson Crusoe as a metaphor, he has based his design work on the use of readily available materials and rejected the trendy work of others. According to Remy, the chest shown here recycles refuse while it incorporates personal recollections, like an American patchwork quilt. He has said: "Each drawer carries its own memories, and these are all jumbled up in your head. So the chest must be just as chaotic." Each chest Remy makes, by its very nature, is different from the other, and few have been made. Remy is a member, if "member" is accurate, of the loosely formed Dutch Droog Design cooperative that arranges for some of its associates' products to be made and publicized. The groups' design work was first shown internationally at the Salone del Mobile (furniture fair) in Milan in 1993 and at a similar event in Cologne, Germany. The début offered the right statement, yet possibly an odd-ball one, at the right time. The press was enthusiastic. Dutch design had languished for what seemed like forever. Possibly not since Rietveld's voice (see p. 48) had sounds from the lowlands been so loud and clear or attained respect from the international design community.

Date: 1991. Materials: Recycled wooden drawers, leather belt. Manufacturer: Tejo Remy, Amsterdam, Netherlands. Photograph: Courtesy The Museum of Modern Art, New York.

Events of 1991:

British architect Norman Foster's Olympia television tower was built, Barcelona.

Premiers: The solo trip by row boat across the northern Pacific Ocean, by Breton-French mariner Gerard d'Aboville; Dow Jones average above 3,000; over 15,000,000 cellular telephones in use worldwide; high-speed trains, Germany.

Music: R.E.M.'s LP "Automatic for the People," Corigliano's "Ghosts of Versailles," Adams's "The Death of Klinghoffer," LP "Unforgettable with Love" with Natalie Cole's and her late father's voices in a duet. Queen's Freddie Mercury died of AIDS.

Film: Holland's *Europa, Europa*, Briseau's *Céline*, Livingston's *Paris Is Burning*, Pascal's *Le petit prince à dit*. The second Chinese film to be nominated for an Academy Award for Best Picture was banned in its native land.

The first woman in 25 years, South African novelist Nadine Gordimer, won the Nobel Prize for literature.

Sanctions, including the 21-year boycott by the Olympic-games committee, were ended against South Africa.

Full-scale war erupted between Iraq and Western Allies, when the U.S. "Stealth" bomber was first put into service.

20-foot tidal wave killed about 120,000 people, Bangladesh.

Marching Neo-Nazi youths worried Europeans, Germany.

The nation of Yugoslavia fell apart.

The much-feared KGB security agency was dissolved, Russia.

Conservative apparatchiks attempted but failed to oust President Mikhail Gorbachev, Russia, and soon the U.S.S.R. was dissolved, creating Russia and other countries.

Thousands of fleeing Albanians were turned back, Italy.

First cholera epidemic in the century broke out, Peru.

Frank Capra, Yves Montand, Robert Maxwell, and Tony Richardson died.

Both former students of industrial design, William Stumpf (b. 1936) studied in Illinois and Wisconsin and Donald Chadwick (b. 1936) in California. In 1970, Stumpf, who was already experienced in office-furniture systems, became vice-president of research at the Herman Miller Furniture Company. After collaborating with others as well as together on furniture for Herman Miller, Stumpf and Chadwick set up a design partnership. Stumpf's "Ergon" office chair of 1966 for Herman Miller had found some success but it was his and Chadwick's "Aeron" chair that was to outshine all their other work, both in its technological prowess and its fame. The polyester-elastomeric mesh for the seat and back, originally used in automobile seats, changes its shape in areas of pressure caused by the sitter and returns when the chair is empty. The height, arms, and back angle can be adjusted to accommodate a number of working conditions, not just sitting at desks or computers. The three available models will accommodate almost all human-body sizes and types. The peculiar-looking suspension system beneath the seat regulates the tilting action. The secret of the chair's success may lie more in the extensive research and marketing funded by Herman Miller than in its designers' acumen, which is considerable. Like a racing car, the "Aeron" is a high-performance machine.

Date: 1992. Materials: Hytrel polymer, polyester, Lycra, die-cast glass-reinforced polyester, aluminum. Manufacturer: Herman Miller Furniture Company, Zeeland, MI, U.S. Photograph: Courtesy Herman Miller Furniture Company.

Events of 1992:

Exposición universal opened, Seville (50,000,000 visitors).

The Australia Group requested its 20-plus industrialist-member nations to end export of a number of pathogens (killer germs) to rogue states; the proposal failed.

Premiers: A Japanese astronaut in space; married couple in space; black woman in space; female priest (Rev. Margaret Philimore of the U.S.) preaching in Canterbury Cathedral, London; 500th anniversary of Christopher Columbus's discovery of America; center for advanced studies in post-communist Eastern Europe (Collegium Budapest), Hungary; longest-reigning monarch in world history (King Sobhuza II of Swaziland); bloodiest conflict since World War II (Serbia's annexation of Bosnia and Hercegovina); Mall of America (the U.S.'s largest shopping mall and world's largest indoor amusement park).

Music: Glass's "The Voyage," Peterson's "The Face of the Night, the Heart of the Dark," Pearl Jam's LP "Ten."

Literature: Keneally's *Schindler's List*, Fernandez's *Dans la main de l'ange*, Böll's *The Safety Net*, Greene's *Monsignor Quixote*, Laurence Olivier's autobiography *Confessions of an Actor*, John Osborne's autobiography *The Life and Crimes of Agatha Christie*.

Theater: Kushner's "Angels in America," a metaphor on social-decay, and Friel's "Dancing at Lughnasa."

Québec's secession was rejected by a plebiscite, Canada.

Princess Diana and Prince Charles separated, and Princess Anne divorced her husband, U.K.

The devastating civil war ended in El Salvador.

The 43rd government since 1945 was formed with the new premier Amintore Fanfani, Italy.

Francis Bacon, Marlene Dietrich, and Emilio Pucci died.

Razor by Ross Lovegrove

Ross Lovegrove (b. 1958) studied at the Polytechnic in Manchester and, until 1982, the Royal College of Art in London. From the early 1990s, he became one of a number of British designers who looked to Europe and elsewhere for the production of their designs. Lovegrove has designed furniture, furnishings, kitchenware, computers, and other products for Knoll, Pottery Barn, frogdesign, Altensteig, Cappellini, and Sony Research. Some of his forms disclose the influence of English sculptor Henry Moore's pebble-smooth masses. More rationalist, the razor here grew out of Lovegrove's affection for new materials and technology. It represents his simple solutions to challenging, hard-to-solve problems. As opposed to the self-conscious products at the end of the century that screamed designer credentials, Lovegrove's plain-Jane razor whispers. The first version of the blade, on the prototype shown here, was a high-alumina-content ceramic, an advanced material like that used by orthodontists. Known as zircon-Y, the substance was coated with a super-thin layer of titanium. Neither material will rust or corrode. If the removable blade does not chip, it will last for a long, long time—a feature not likely to attract razor-blade manufacturers whose main interests lie in obsolescence.

Date: 1993. Materials: Titanium-plated zircon-Y (blade), injection-molded acrylic (handle) (prototype shown here). Manufacturer of prototype: M.L. Laboratories, U.K. Photograph: © The Museum of Modern Art, New York; exhibition and catalogue "Mutant Materials in Contemporary Design," 1995.

Events of 1993:

"Design: Mirroir du siècle" exhibition was organized at the Grand Palais, Paris.

"Shock art" appeared at the 45th Venice Biennial where artist Louise Bourgeois declared, "I want to bother people."

Premiers: A black female recipient of the Nobel Prize; female U.S. attorney general; females in combat roles, U.S; fast Internet system ("T3"), replacing the previous configuration of the U.S. Defense Department and financed by U.S. National Science Foundation, Washington; H.I.D. (high-intensity discharge) auto headlamp, by BMW, Germany; reopening of Max Planck Society (successor to the Kaiser Wilhelm Institute), Germany; paid public admission to Buckingham Palace, London; ratification of the Treaty on European Union ("Maastricht Treaty"), aimed at creating a single European Community currency; artillery-shell damage to the Ottoman bridge, unscathed since the 16th century, Mostar, Bosnia.

Music: k.d. lang's song "Constant Craving," Pearl Jam's LP "Vs," Sting's "If I Ever Lose My Faith in You," Nanes's "Holocaust Symphony."

Film: Campion's *The Piano*, Kaige's *Farewell My Concubine*, Ephron's *Sleepless in Seattle*.

Hal Prince's musical play "Kiss of the Spider Woman" opened on Broadway, New York.

War continued with Allied forces' bombing of Iraqi sites.

Mafia bombs killed 5 people and damaged the Uffizi Gallery, an apartment complex, two of the oldest churches in Rome, and an historic center in Milan, Italy.

The World Trade Center was bombed by Islamic terrorists, killing 6 people and injuring more than 1,000, New York.

Kobo Abé, Frederico Fellini, and Audrey Hepburn died.

Yet another Italian architect-designer to have studied at the Politecnico in Milan, Italy, Antonio Citterio (b. 1950) began a career in industrial design in 1967 and after the Politecnico set up a studio in 1972 with Paolo Nava with whom he collaborated until 1981. A very busy designer, Citterio frequently works with his American wife Terry Dwan (b. 1957) and occasionally G. Oliver Löw. His client list, which is extensive, includes Kartell, the distinguished Italian plastics manufacturer (see p. 142). For the Kartell cabinets shown here, Citterio chose a different kind of plastic to the one Kartell normally uses. Even though ABS is the material of the opaque slate-colored version, a translucent high-performance thermoplastic technopolymer for the other colors offers scratch resistance and creates a ghost-like effect. In the 1970s, plastics were produced in garish colors and became shabby in time. On the contrary, the technopolymer of Citterio's cabinets comes in a range of four attractive, though vibrant, colors and is highly durable. The aluminum frame of the "Mobil" offers stability, and the contrast of the strength of metal and gossamer lightness of the synthetic material is highly agreeable. Citterio has the ability to create furnishings that are unusual but not so different as to be off-putting. The success of the "Mobil" spawned a number of inexpensive imitations.

Date: 1994. Materials: Technopolymeric thermoplastic for four colors and ABS for the slate color, chromium-plated or painted aluminum. Manufacturer: Kartell S.p.A., Noviglio (MI), Italy. Photograph: Courtesy Kartell S.p.A.

Events of 1994:

Premiers: Traffic through the Chunnel, the tunnel between England and France under the English Channel, accommodating train travel at 90 to 180 m.p.h., costing $15 billion and employing 15,000 workers for seven years; diplomatic relations between Israel and the Vatican; all-female America's Cup sailing team; conclusive evidence of black holes, through the use of the outerspace Hubble telescope; American Center, by American architect Frank O. Gehry, with critical comment that it looked "too French," Paris; discovery of a meat-eating dinosaur, by U.S. and Mongolian scientists, Gobi Dessert; official NATO offensive war action (first in 45 years), Bosnia; black South African head of state (Nelson Mandela); a Russian on an American space shuttle.

Music: LP "Chant" by Benedictine Monks of Santo Domingo, Rolling Stones' LP "Voodoo Lounge," Dion's song "The Power of Love," Snoop Doggy Dog's song "What's My Name?," R.E.M.'s LP "Monster," Schnittke's "Symphony No. 7."

Film: Newell's *Four Weddings and a Funeral* and Tarantino's *Pulp Fiction*.

Andrew Lloyd Webber's musical play "Sunset Boulevard" opened on Broadway, New York.

A major earthquake (6.6 Richter) struck, with $7 billion in damages and 34 people dead, Los Angeles.

Carlos the Jackal, the world's most-wanted terrorist, was captured and incarcerated in La Santé prison, Paris.

Revengeful Phoolan Devi ("the Bandit Queen") was released from jail after 11 years, India.

After a 27-year exile, PLO leader Yasser Arafat crossed the Egyptian border and entered the Gaza Strip.

CIA masterspy Aldrich Ames was given a life sentence, U.S.

Eugène Ionesco and Jacqueline Kennedy Onassis died.

Marcel Wanders (b. 1963) graduated from the college of fine arts in Arnhem, Holland, in 1988. In the following year, he produced his first design, a lamp commissioned by the PTT (Dutch postal service). For the assignment, he used ordinary, plain cloth lampshades, several stacked one over the other. When he realized that critics were calling the simple solution a "ready-made," he became upset and asserted that it was rather an academic concept. But there are those who may think there is no difference. Next, in 1995, came an invitation to participate in the so-called Dry Project sponsored by the Droog Design Foundation and a chance to realize his aspiration to "make things with love" for what, he says, has become a technically oriented society. The idea that began with macramé took the form of a chair that "should be usable and durable and not lose its value with age," according to Wanders. To build the chair, rope made of carbon fiber within an aramide-braided sleeve was knotted, like macramé, into a limp, unrecognizable chair form. It was then dipped into an epoxy solution, hung in a skeletal frame, pulled out at eight corners to form the shape of a chair, and dried in a special room under high heat. The marriage of handicraft and high technology begot a deceptively sturdy seat almost as strong as iron. Considering that the chair is essentially a handmade object, a surprising number have been produced.

Date: 1995. Materials: Carbon fiber, aramide, epoxy. Manufacturer: Marcel Wanders, Amsterdam, Netherlands (from 1995); Cappellini, Milan, Italy (from 1998). Photograph: Courtesy Marcel Wanders.

Events of 1995:

Russian astronaut Valeri Polyakov spent a record 438 days in space. NASA's spacecraft *Discovery* rendezvoused with Russian space station *Mir*. First female pilot (Eileen Collins) flew the Space Shuttle.

Directory of human genomes, the encyclopedia of the human genetic code, was published by the journal *Nature*.

The Wolfsonian design museum was founded by Michael Wolfson, Miami Beach, Florida, U.S.

Premiers: The "100% Design" fair, London; Internet Explorer 2.0, by Microsoft; broadcast of commercial 24-hour radio (Radio HK), an Internet-only radio station; Vatican Web site (http://www.vatican.va/); WWW (Worldwide Web search engines); solution of Fermat's last theorem, thought to be impossible; Sony PlayStation hardware for computer games, becoming far more successful than competitors'.

U.S. President Bill Clinton first offered "temporary use" of ground troops to NATO in Bosnia Hercegovina. A tribunal charged Serb leader Karadzin with genocide.

NATO held its first military exercises in West Germany.

New Zealand sailboat *Black Magic 1* swept the America's Cup final, off San Diego, California.

Jacques Chirac became the president of France.

Police officers rejected routinely carrying weapons, U.K.

Collapse of Barings Bank, one of the U.K.'s oldest investment houses, was blamed on rogue dealer Nick Leeson, a U.K. citizen based in Sinapore.

4,000 people died in an earthquake in Kobe, Japan.

Sarin nerve gas was released by Buddhist cult members, killing a dozen people and injuring thousands, Tokyo, Japan.

Ginger Rogers, Donald Pleasance, and Harold Wilson died.

A native of London, Tom Dixon (b. 1959) studied at the Chelsea School of Art until 1978 when he became a sculptor. In 1983, he completed his first interior-design commission and, the next year, began welding scrap metal on-stage as performance art. For a time, he collaborated with André Dubreuil, a designer who specialized in wrought iron furniture and furnishings. From 1986–87, Dixon also produced furniture in steel and other one-of-a-kind pieces. In 1987, he set up the production workshop Dixon PID while collaborating with others. His "S" chair of 1988, at first made with rubber strips wound around a steel-rod frame, was eventually produced by Cappellini in rush fiber. Like other British designers, Dixon sought associations with manufacturers outside the country, particularly in Italy, due to the absence in the U.K. of sophisticated producers of furniture who were interested in cutting-edge design. In 1996, he further countered by founding Eurolounge to manage the production of his products like the "Jack" lamp shown here. Made in a thermoformed plastic material, the arrangement of the lamp's protrusions facilitates multiple configurations—whether on a floor or a desk or from a ceiling. When a flat top is added, the lamp becomes an illuminated stool or table and satisfies Dixon's devotion to multi-purpose. The pigmentation of the polyethylene material was never before applied to furniture production.

Date: 1996. Materials: Thermoformed polyethylene. Manufacturer: Eurolounge Ltd., London. Photograph: Courtesy Eurolounge.

Events of 1996:

*Wallpaper** magazine began publication, with Tyler Brûlé as editor-in-chief, London, and in 1997 was purchased by Time Warner.

Premiers: The oldest person to fly in space to date (Story Musgrave); spacecraft to land on another planet (Venus), by the Soviet *Venus III*.

Reacting to the efforts of the members of the cult Aum Shinrikyo in Japan and American hate-monger Larry Wayne Harris in America, the U.S. Congress signed into law a bill that criminalized the unauthorized use of pestilential germs and imposed stringent rules on their transfer, but few other countries and groups followed.

Research into the effects of animals and humans living in outerspace was greatly increased.

Restrictions on Internet use around the world: China required users and ISPs to register with the police. Germany cut off access to some news groups carried on Compuserve. Saudi Arabia confined Internet access to universities and hospitals. Singapore required political and religious content providers to register with the state. New Zealand classified computer disks as "publications" that can be censored and seized.

Film: *Secrets & Lies* received the Palm d'or, Cannes.

Malaysian prime minister Mahathir Mohamad, PLO leader Yasser Arafat, and Philippine president Fidel Rhamos met for ten minutes in an online interactive "chat session."

The U.K. and France became alarmed by BSE (Bovine Spongiform Encephalopathy), a deadly disease in cattle.

118 people died in a train crash, Pakistan, and 230 people died in the airplane crash of TWA flight "800," off Long Island, New York. Gianni Versace was murdered in Miami. Ella Fitzgerald also died.

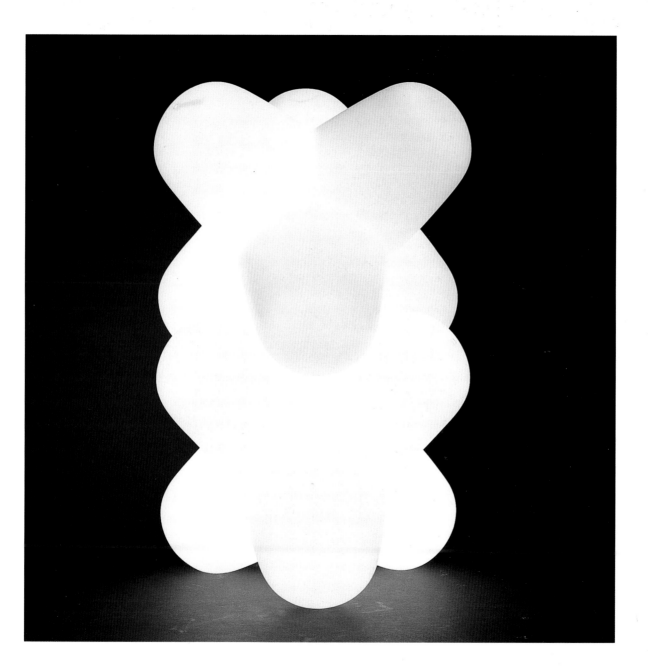

"Focus" mixing bowl by Hansjerg Maier-Aichen

Hansjerg Maier-Aichen (b. 1940) studied interior design and painting in Germany and received a master's degree from the University of Chicago. After several teaching posts, he became managing director of artipresent GmbH in Germany. Equally adept at both the business of design and the aesthetics of design, Maier-Aichen created the AUTHENTICS brand in 1980 within the artipresent organization. Its products achieved a high-profile presence in the tabletop market in the 1990s when the firm introduced a kitchen and bathware line in recyclable polypropylene. The collection acknowledged plastics as legitimate materials with their own respectable value and not as imitations of traditional materials, like wood or metal. The luminescence of planet-respecting recyclable polypropylene caught the imagination of the public. The success of the line was further propelled by Maier-Aichen himself and his stable of free-wheeling freelance designers who created clean, imaginative objects. Some, like trash bins and the bowl (shown here), could be put to serious use. Others, like cups that separated toothbrushes from one another, were clever. And items, like wall knobs that held ring-shaped soap, were amusing. Glassware and furniture were eventually added to the line, but the translucent nature of the polypropylene in inviting colors is the essence of the collection's special character. The abounding number of imitators has validated what was a propitious, good idea.

Date: 1997. Materials: Recyclable polypropylene. Manufacturer: artipresent GmbH, Holzgerlingen, Germany. Photograph: Courtesy AUTHENTICS. Photo: Markus Richter Stuttgart.

Events of 1997:

An in-production V8-engine motorcycle was introduced by Morbidelli, Italy.

The Oreo cookie was certified kosher.

Texas Instruments' TIRIS technology made pay-at-the-pump gasoline (petrol) purchases possible through use of Mobil's Speedpass.

Swiss banks planned the first payment to holocaust victims or families.

Eric Clapton received a Grammy Award for best song, "Change the World," and Celine Dion for best album.

Dow Jones industrial average of 30 blue-chip stocks fell 554 points, the greatest drop ever; then rose 337 points, the highest-ever single increase, U.S.

The first fossilized skin of 70- to 90,000,000-year-old unhatched dinosaurs was found in embryos, Argentina.

Tony Blair was elected the prime minister of the U.K.

Eric Schulman published the dubiously contributory *History of the Universe in 200 Words or Less*.

DVD (digital video disk) technology was introduced.

"Internet Official Protocol Standards" were established.

World's biggest to date: circulation of a daily newspaper, *Yomiuri Shimbun* at 14,600,000, Japan; revenue of a film, *Titanic* at $582,000,000; sales of a music album, Jackson's "Thriller" at 25,000,000 copies; run of a Broadway show, Webber's "Cats"; number of personal computers in the U.S., at 91,500,000.

E.A.S. (Emergency Alert System), a public disaster announcement system, employed by radio-broadcast station, replacing the aging E.B.S.(Emergency Broadcasting System).

A court convicted Carlos the Jackal of murder, Paris.

11,700,000 people had died from AIDS worldwide. Diana—Princess of Wales, Linda McCartney, Pol Pot, Alan Shepard, and Frank Sinatra also died.

The cabinet of Apple Computer's "iMac" was principally in the hands of Jonathan Ive (b. 1967), Apple's head of design since 1996. He had studied industrial design in England before being introduced to Apple while working at Tangerine, a London design studio he co-founded in 1990. His first design as an Apple employee was the "Newton MessagePad 110" of 1992. Numerous other assignments followed including the "20th Anniversary Macintosh" of 1993–96. Two levels below Steven Jobs who became the returning savior of Apple in 1997, Ive vowed that Apple's next products would be "targeted to individuals rather than large unfocused groups. . . and be the most exciting and meaningful designs that Apple has ever delivered." The pledge was realized as the "iMac." Fanatical attention was paid to every minute aspect of the integrated body, the compatible keyboard, and the pill-box mouse—machinery intended to put households in touch with the Worldwide Web. The translucent shell revealed a fuzzy picture of the powerful workings inside. Even the mouse's intestines could be seen. The tangle of wires found at the back end of previous beige-bodied computers was banished. The "iMac" was a far cry from Jobs' first born, the "Apple I" of 1975, and the formality of Hartmut Esslinger's "Apple SE" of 1987 (see p. 188). A century later, the first electric typewriter, the Blickensderfer of 1901, had miraculously metamorphosed into the little "blue egg."

Date: 1998. Materials: Plastic, metal, glass, halogravure printing. Manufacturer: Apple Computer, Inc., Cupertino, CA. Photograph: Courtesy Apple Computer Company, Inc. Photo: Terry Hefferman.

Events of 1998:

The total population of the world was about 5.8 billion.

Stade de France (sports stadium) was built in time for Le Coup de Monde event, St. Denis, France. (France won the tournament.)

Premiers: The International Space Station (first component); 175-m./h. motorcycle ("MV Augusta F4"), Italy; oldest person in space (John Glenn, also the first American to orbit the Earth, in 1962); the cruise ship *Grand Princess* (twice the size of the *Titanic*); 14-day boat trip from Japan to U.S., by French sailor Bruno Peyron; discovery of Sterkfontein skeleton (missing link between apes and humans); worst train accident in 50 years (almost 100 died), Germany; first line of the Métro subway in 63 years, Paris; spacecraft with ion-propulsion, by NASA; improved large-vocabulary voice recognition for computers, by IBM; conversion of the Internet into a non-profit international organization.

Rolls-Royce auto name and its "Spirit of Ecstasy" emblem were purchased by BMW, and the Bentley marque by Volkswagen, both German auto firms.

Swatch, known primarily for its watches, and Mercedes introduced the "Smart" car with color and parts changeable within two hours by the owner.

HDTV (high-definition television) sets cost $8,000–10,000, with limited programming available, U.S.

Corot's *The Sèvres Road* was stolen from the Louvre in Paris, recalling the theft of the *Mona Lisa* in 1911.

Film: *Godzilla*, *Armageddon* with Bruce Willis, Steven Spielberg's *Saving Private Ryan*, *Antz*.

International corporations merged into huge entities.

Hurricane Mitch erased entire villages in Central America.

Swiss banks were forced to pay $1.25 billion to Holocaust victims or their heirs.

A wave of teenage murders by other teenagers swept the U.S.

Nicholas II of Russia, his family, and others were buried in Moscow, 80 years to the day after they were murdered.

Robotic vacuum cleaner by Ljunggren, Ljunggren, & Haegermark

In 1901, British bridge engineer Hubert Cecil Booth patented a suction vacuum cleaner; it was the size of a modern domestic refrigerator. The principle was refined further and further until 1912 when Axel Wenner-Gren (1881–1961) perfected a 14-kg model, the "Lux." A decade later, in 1921, he founded Electrolux and produced the first vacuum cleaner for ordinary households. Still, it was far from offering the kind of convenience that a robotic vacuum might. Yet, in 1991, an expensive, unwieldy robot was introduced in the U.S. for industrial use. "For a long time I've had this vision of a robotic cleaner for ordinary people living in ordinary houses," said Per Ljunggren (b. 1957), the eventual leader of a project at Electrolux for just such a device. The transformation of idea into reality began with a core research team, later joined by other experts. Technology and design were to be equal partners. Mechanical engineer Anders Haegermark (b. 1964) worked closely with industrial designer Inese Ljunggren (b. 1954), who has declared her role: "As the industrial designer, I represent the user." A prototype was modeled in clay with the bumper also serving as handle. The air-evacuation gills drew comparisons by the team to the prehistoric trilobite. By 1999, a limited number of models had been made by means of an ultraviolet beam that hardened a photopolymeric material layer by layer. The sample had the same finish and feel of the final canisters.

Date: 1999. Materials: Photopolymeric material, metal, electronics (prototype shown here). Manufacturer: Electrolux, Stockholm, Sweden. Photograph: Courtesy The Eureka Company, Bloomington, IL, U.S.

Events of 1999:

Compaq became the world's largest maker of personal computers.

Premiers: A manned balloon traveling completely around the world (*Breitling Orbiter 3*); the euro as Europe's standard currency, with the first phase by credit cards and bank checks only; American high-speed train (between Boston and New York); more speech commands than any other auto (Jaguar "S-Type"); legalized sexual prostitution in Switzerland; Japanese heart transplant (first in 30 years in Japan); Dow Jones stock average at 11,000 points by May and still rising.

Film: Kapur's *Elizabeth*, Madden's *Shakespeare in Love*, Benigni's *Life Is Beautiful*, Salles's *Central Station*, Ephron's *You've Got Mail*, *The Truman Show*, *Pleasantville*, Lucas's *Star Wars—Episode I: The Phantom Menace*.

Michel Houellébecq's book *Les particule élémentaires* derided French "baby boomer" governmental leaders.

After the first year of the only legally sanctioned assisted suicides in the world, 15 people had died in Oregon, U.S.

A Danish scientist and others slowed the speed of light to 38 m./h. by freezing atoms to almost absolute zero, thus making almost no movement possible, U.S.

Record snowfall caused avalanches and numerous deaths in the Alps of Europe and elsewhere. 300 m./h. cyclones swept through the Great Plains states reeking extensive damage and causing deaths, U.S.

A wave of corporate megamergers proliferated worldwide.

NATO forces attacked Yugoslavian army to thwart atrocities against Albanians and others in Kosovo.

200,000 abortions were registered, France.

AIDS virus was identified as having originated with a certain subspecies of chimpanzee in Central West Africa, possibly carried for 100,000 years.

U.S. President Bill Clinton was impeached for perjury but remained in office.

Dirk Bogarde, Sir Yehudi Menuhin, and Oliver Reed died.

Acknowledgements

The patient assistance of the following colleagues
and friends and the kindness of strangers made this
volume possible:

Alberto Alessi

Cinzia Anguissola d'Altoé

Linda d'Anjou, Le Musée des Arts Décoratifs de Montréal

Brice d'Antras

Paola Antonelli, The Museum of Modern Art, New York

James Benjamin

Georg Christof Bertsch

Stéphane and Catherine de Beyrie, Galerie de Beyrie

Jill Bloomer, Cooper-Hewitt National Museum of Design

Dr. Claire Bonney

Gianluca Borgesi, Zanotta

Frederick R. Brandt, Virginia Museum of Fine Arts

Matthias Brüllmann

Christine Butt

Bruce Buursma, Herman Miller Furniture Company

M.P. Carey, Time Will Tell

Diane Charbonneau, Le Musée des Arts Décoratifs de
    Montréal

Christie's Images

Isabella Colombo, Cassina S.p.A.

William Costa, Luceplan

Morrison Cousins, Tupperware World Headquarters

Deanna Cross, The Metropolitan Museum of Art

Eames Demetrios, Eames Office

Isabelle Denamur

Valentina Di Stephano, Solari Udine

Maria Helena Estrada

Fabrio Ferri, Necchi S.p.A.

Patrick Grandy, Zippo Manufacturing Company

Eliza Grassy, National Housewares Company

Tod Gustafson, George Eastman House

Jean Hines, Pratt Institute Library

Arlene Hirst, *Metropolitan Home*

Florian Hufnagl, Die Neue Sammlung

Steven Jaffe

Donald Kalec

JoAnn Klein, 3M

Guto Lacaz

Milena Lamarová, Umeleckoprumyslové Muzeum v Praze

Olavi Lindén, Fiskars Consumer Oy Ab

Yrsa Lindström, Fiskars Consumer Oy Ab

Katherine Luedke, The Eureka Company

Ivan Luini, I.L. Euro

Claude Lichtenstein, Museum für Gestaltung Zürich

Randell L. Makinson

Janice Madhu, George Eastman House

Linda Meabon, Zippo Manufacturing Company

Torben Holme Nielsen, Fritz Hansen A/S

Andreas Nutz, Vitra Design Museum

Linda O'Keefe, *Metropolitan Home*

Todd Olson, Cooper-Hewitt National Museum of Design

Dr. Eugenio Pacchioli, Archivio Storico Olivetti

Noralyn Pease

Howell W. Perkins, Virginia Museum of Fine Arts

Stefano Perrone, Pesce, Ltd.

Gaetano Pesce

Joe Pollock, Wurlitzer Jukebox Company

Bill Porter

Sarah Robins, The Museum of Modern Art, New York

Mary Rogers, Cuisinart Co.

Annette Ruggerio, Cassina U.S.

Gad Sassower, Decodence
Stuart Schneider
Bonnie Schwartz, *I.D.*
Sheryl Shade
Judy Shuster, 3M
Josef Straßer, Die Neue Sammlung
Dan Thelander, Bruno Mathsson International AB
Philippe Thonet, Gebrüder Thonet GmbH
Peter van der Jagt
John C. Waddell
Marcel Wanders
Wolfgang Weingart
Katherine Wildt
Bruce N. Wright

Reference Works

The following publications were valuable in the preparation of this volume:

Catherine Armijon et al., *L'Art de Vivre: Decorative Arts and Design in France 1789–1989*, New York: The Vendome Press and Cooper-Hewitt Museum, 1989.

Eric Baker, text by Jane Martin, *Great Inventions, Good Intentions*, San Francisco: Chronicle Books, 1990.

Arlette Barré-Despond, ed., *Dictionnaire internationale des arts appliqués et du design*, Paris: Éditions du Regard, 1996.

Stephen Bayley, ed., *In Good Shape: Style in Industrial Products, 1900 to 1960*, New York: Van Nostrand Reinhold Co., 1979.

Christoph Bignens, "Hermes-Baby Schreibmachine," *Unbekannt-Vertraut*, Zürich: Museum für Gestaltung Zürich, 1987.

Andrea Branzi and Michele De Lucchi, *Il Design italiano degli Anni '50*, Milan: Centrokappa and R.D.E., Richerche Design Editrice, 1985.

Bernhard E. Bürdek, *The Apple Macintosh*, Frankfurt am Main: Verlag form GmbH, 1997.

Mel Byars, *The Design Encyclopedia*, London: Laurence King, 1994.

Frederick R. Brandt, *Late 19th and Early 20th Century Decorative Arts: The Sydney and Frances Lewis Collection in the Virginia Museum of Fine Arts*, Richmond, VA: Virginia Museum of Fine Arts, 1985.

Robert Judson Clark, *The Arts and Crafts Movement in America 1876–1916*, Princeton, NJ: Princeton University Press, 1972.

Tessa Clark, ed., *Bakelite Style*, London: Quintet Publishing Limited, 1997.

Clifton Daniel, ed., *Chronicle of the 20th Century*, New York: Dorling Kindersley Publishing, Inc., 1995.

Jay Doblin, *One Hundred Great Product Designs*, New York: Van Nostrand Reinhold Co., 1970.

Domus journal, Milan.

Magdalena Droste, *The Bauhaus-Light by Carl Jacob Jucker and Wilhelm Wagenfeld*, Frankfurt am Main: Verlag form GmbH, 1977.

Martin Eidelberg, ed. *Design 1935–65: What Modern Was, Selections from the Liliane and David M. Stewart Collection*, New York: Harry N. Abrams, Inc., Publishers, and Le Musée des Arts Décoratifs de Montréal, 1991.

Allyn Freeman and Bob Golden, *Why Didn't I Think of That?: Bizarre Origins of Ingenious Inventions We Couldn't Live Without*, New York: John Wiley & Sons, Inc., 1997.

Lorraine Glennon, ed., *Our Times: The Illustrated History of the 20th Century*, Atlanta: Turner Publishing, Inc., 1995.

Henry Gostony and Stuart Schneider, *The Incredible Ball Point Pen*, Atglen, PA: Schiffer Publishing Ltd., 1998.

Alfonso Grassi and Anty Pansera, *Atlante del design italiano 1940/1980*, Milan: Gruppo Editoriale Fabbri, 1980.

Bernard Grun, *The Timelines of History*, 3d ed., New York: Simon & Shuster, 1991.

Alexander Hellemans and Bryan Bunch, *The Timetables of Science: A Chronology of the Most Important People and Events in the History of Science*, New York: Simon & Shuster, 1988.

Kathryn B. Hiesinger and George H. Marcus, *Design Since 1945*, Philadelphia: Philadelphia Museum of Art, 1983.

Kathryn B. Hiesinger and George H. Marcus, *Landmarks of Twentieth-Century Design*, New York: Abbeville Press, Publishers, 1993.

Florian Hufnagl, ed., A *Century of Design: Insights, Outlook on a Museum of Tomorrow*, Stuttgart: Arnoldsche, 1996.

I.D. magazine, New York.

Guy Julier, *New Spanish Design*, London: Thames and Hudson Ltd, 1991.

Randell L. Makinson, *Greene and Greene: The Passion and the Legacy*, Salt Lake City, UT: Gibbs Smith Publishers, 1998.

Randell L. Makinson, *Greene and Greene: The Blacker House*, Salt Lake City, UT: Gibbs Smith Publishers [in press].

Vincent A. Masucci, *The Classics of Modern Furniture*, 3d ed., New York: Palazzetti, 1992.

François Mathey, director, and Arlette Barré-Despond, coordinator and iconographer, *À bonheur des formes: design français 1945–1992*, Paris: Éditions du Regard, 1992.

Victoria Kasuba Matranga with Karen Kohn, *A Celebration of Twentieth-Century Housewares*, Rosemont, IL: National Housewares Manufacturers Association, 1997.

Catherine McDermott, *Design Museum Book of 20th Century Design*, Woodstock, NY: The Overlook Press, 1998.

David Revere McFadden, ed., *Scandinavian Modern Design 1880–1980*, New York: Harry N. Abrams, Inc., Publishers, 1982.

Craig Miller, *Modern Design 1890–1990*, New York: Harry N. Abrams, Inc., Publishing, and The Metropolitan Museum of Art, 1990.

Renato Minetto, ed., *Design italiano nei musei moderna*, Rome: Galleria Nationale d'Arte Moderna, 1988.

Jennifer Hawkins Opie, *Scandinavia: Ceramics and Glass in the Twentieth Century*, New York: Rizzoli, 1989.

Derek E. Ostergard and Nina Stritzler, eds., *The Brilliance* of Swedish Glass, 1918–1939: An Alliance of Art and Industry, New Haven, CT: Yale University Press, 1996.

Ottagono journal, Milan.

Charles Panati, The Browser's Book of Beginnings, Boston: Houghton MIfflin Company, 1984.

Charles Panati, Extraordinary Origins of Everyday Things, New York: Harper & Row, Publisher, 1987.

David Poore, Zippo: The Great American Lighter, Atglen, PA: Schiffer Publishing Ltd., 1997.

Margo Rouard and Françoise Jollant Kneebone, directors of the work, Design Français 1960–1990, Paris: A.P.C.I./ Centre Georges Pompidou, 1988.

Richard Sexton, American Style: Classic Product Design from Airstream to Zippo, San Francisco: Chronicle Books, 1987.

Penny Sparke, Design in Italy: 1870 to the Present, London: John Calmann and King Ltd, 1988.

Penny Sparke, A Century of Design: Design Pioneers of the 20th Century, London: Mitchell Beazley, 1998.

Joan Wessel with Nada Westerman, American Design Classics, New York: Design Publications, Inc., 1985.

Hans Wichmann, Industrial Design, Unikate, Serienerzeugnisse: Die Neue Sammlung, Ein neuer Museumstyp des 20. Jahrhunderts, Munich: Prestel-Verlag, 1985.

Hans Wichmann, Italien Design 1945 bis heuter, Munich: Die Neue Sammlung, 1988.

Hans Wichmann, Donationen und Neuerwerbungen 1986/87, Munich: Die Neue Sammlung, 1989.

Hans Wichmann, Donationen und Neuerwerbungen 1988/89: Industrial Design, Unikate, Serienerzeugnisse, Munich, Die Neue Sammlung, 1990.

Hans Wichmann, Die Realisation eines neuen Museumstyps: Die Neue Sammlung, Bilanz 1980/90, Basel: Birkhäuser Verlag, 1990.

Nancy Winters, Man Flies: The Story of Alberto Santos-Dumont, Master of the Balloon, Conqueror of the Air, London: Bloomsbury Publishing Ltd, 1997.